S0-AKJ-167

COMPUTER-ASSISTED INSTRUCTION IN CHEMISTRY

(IN TWO PARTS)

Part A: General Approach

COMPUTERS IN
CHEMISTRY AND INSTRUMENTATION

edited by

JAMES S. MATTSON HARRY B. MARK, JR. HUBERT C. MACDONALD, JR.

Volume 1: Computer Fundamentals for Chemists

Volume 2: Electrochemistry: Calculations, Simulation, and Instrumentation

Volume 3: Spectroscopy and Kinetics

Volume 4: Computer-Assisted Instruction in Chemistry (in two parts). Part A: General Approach. Part B: Applications

Volume 5: Laboratory Systems and Spectroscopy (in preparation)

COMPUTER-ASSISTED INSTRUCTION IN CHEMISTRY

(IN TWO PARTS)

Part A: General Approach

EDITED BY

James S. Mattson

Division of Chemical Oceanography
Rosenstiel School of Marine and Atmospheric Sciences
University of Miami
Miami, Florida

Harry B. Mark, Jr.

Department of Chemistry
University of Cincinnati
Cincinnati, Ohio

Hubert C. MacDonald, Jr.

Koppers Company, Inc.
Monroeville, Pennsylvania

MARCEL DEKKER, INC. New York 1974

J.C.

MARCEL DEKKER, INC.
305 East 45th Street, New York, New York 10017

LIBRARY OF CONGRESS CATALOG CARD NUMBER: 73-89669
ISBN: 0-8247-6103-0

Printed in the United States of America

INTRODUCTION TO THE SERIES

In the past decade, computer technology and design (both analog and digital) and the development of low cost linear and digital "integrated circuitry" have advanced at an almost unbelievable rate. Thus, computers and quantitative electronic circuitry are now readily available to chemists, physicists, and other scientific groups interested in instrument design. To quote a recent statement of a colleague, "the computer and integrated circuitry are revolutionizing measurement and instrumentation in science." In general, the chemist is just beginning to realize and understand the potential of computer applications to chemical research and quantitative measurement. The basic applications are in the areas of data acquisition and reduction, simulation, and instrumentation (on-line data processing and experimental control in and/or optimization in real time).

At present, a serious time lag exists between the development of electronic computer technology and the practice or application in the physical sciences. Thus, this series aims to bridge this communication gap by presenting comprehensive and instructive chapters on various aspects of the field written by outstanding researchers. By this means, the experience and expertise of these scientists are made available for study and discussion.

It is intended that these volumes will contain articles covering a wide variety of topics written for the nonspecialist but still retaining a scholarly level of treatment. As the series was conceived it was hoped that each volume (with the exception of Volume 1 which is an introductory discussion of basic principles and applications) would be devoted to one subject; for example, electrochemistry, spectroscopy, on-line analytical service systems. This format will be followed wherever possible. It soon became evident, however, that to delay publication of completed manuscripts while waiting to obtain a volume dealing with a single subject would be unfair to not only the authors but, more important, the intended audience. Thus, priority has been given to speed of publication lest the material become dated while awaiting publication. Therefore, some volumes will contain mixed topics.

The editors have also decided that submitted as well as the usual invited contributions will be published in the series. Thus, scientists who have recent developments and advances of potential interest should submit detailed outlines of their proposed contribution to one of the editors for consideration concerning suitability for publication. The articles should be imaginative, critical, and comprehensive survey topics in the field and/or other fields, and which are written on a high level, that is, satisfying to specialists and nonspecialists alike. Parts of programs can be used in the text to illustrate special procedures and concepts, but, in general, we do not plan to reproduce complete programs themselves, as much of this material is either routine or represents the particular personality of either the author or his computer.

The Editors

PREFACE

As chemistry is an important part of most other scientific fields, such as biology, medicine, geology, pharmacy, environmental sciences, engineering, and materials, university and college students majoring in these areas are generally required to take one or more years of chemistry courses. Furthermore, large numbers of liberal arts majors elect to take chemistry as the required physical science in their program. Thus, in large universities as many as five thousand students per year will take freshman chemistry and hundreds will continue on into organic, analytical, and physical chemistry. In the small colleges, although the total numbers are smaller, the size of each class and laboratory is generally just as large with respect to student/teacher ratio. As the prospect for increased enrollment grows, while that for significant additional staff positions decreases in general, this ratio will become even more unfavorable in the next ten years in most institutions. Many chemistry departments are beginning to explore the possibilities of computer-assisted instruction (CAI) techniques and methods in an effort to improve the quality of instruction not only in large classes, but also in small classes, to reduce classroom and laboratory costs, reduce time-consuming teacher tasks, such as grading, and to provide special instruction to certain groups of students.

Recently, there have been several conferences, articles in the Journal of Chemical Education, etc., which have presented the results of the initial efforts in CAI in chemistry by numerous groups. It was felt that it would be useful and timely to assemble the various methods and opinions on applications of CAI in chemistry in one place, as there are very large numbers of teachers and chemistry departments throughout the country who are considering employing CAI and would benefit from the experiences of pioneers in the area. Thus, we have collected chapters on various aspects of CAI by chemistry teachers from both large and small institutions who have been actively developing and applying CAI techniques. The readers will notice that there is considerable overlap and duplication in the material presented in the different chapters in this volume. We did this purposely as we felt that CAI in chemistry is a developing and evolving teaching technique and that there are no proven methods. We wanted different authors to express their opinions, ideas, approaches, etc., on the same objects and classroom needs. In this way, the reader can evaluate these diverse opinions and experiences in light of his particular situation and, hopefully, arrive at the best approach to the solution of his teaching problems.

This volume is organized into two parts, A and B, for convenience. Part B is subdivided further into two sections. However, the reader should note that, in most cases, portions of chapters in one category also overlap or fall into other categories.

Chapter 1 of Part A serves as a general guide to the philosophy, approaches, and applications of CAI in general. It also serves as an overview to the material contained in the rest of the volume. Part A also contains five chapters which discuss general techniques and applications. Chapter 2 is a detailed presentation of computer-assisted instruction and computer-augmented learning techniques that have been developed and applied to the curriculum at practically all levels at the University of Pittsburgh. Chapter 3 deals with the basic concepts of analog and hybrid computers. Although this chapter is not principally concerned with teaching applications, we felt analog and hybrid computers and computation have great potential for chemistry CAI in simulation of systems and instrumentation. These have been used extensively in engineering sciences and, thus, we felt that it would be very worthwhile including a comprehensive discussion of basic principles in this volume to introduce the reader to the subject. Chapter 4 presents the concept of video projection of teletype output to enable the instructor to use the computer on-line in the classroom. As the volume of chemical literature is expanding at an enormous rate, it is important that computerized information storage and retrieval practice, as discussed in Chapter 5, be introduced to the students (and instructors). One of the questions that readers who are contemplating introducing CAI into their curriculum may have is cost. Chapter 6 gives a brief resume of a cost estimate for various configurations of CAI systems.

Part B, Section 1, which also discusses general approaches, is specifically aimed at presenting detailed discussions of the techniques, programs, and experiences in special applications or courses. Chapter 1 deals with the use of APL language in CAI, and Chapter 2 discusses the use of the PLATO system at the University of Illinois in organic chemistry courses. Chapter 3 is concerned with CAI in physical chemistry courses.

Part B, Section 2 deals with special applications and techniques. Chapter 4 discusses detailed aspects of the use of the computer in generating tests, and Chapter 5 is concerned with the use of a time-shared computer system in CAI. Computer simulation of unknowns is covered in Chapter 6, the use of "canned" programs is discussed in Chapter 7, and Chapter 8 deals with techniques of computerized homework preparation and grading.

The Editors wish to gratefully acknowledge the tremendous editorial help of Dr. Thomas A. Atkinson, Department of Chemistry, Michigan State University and Dr. Richard L. Ellis, Department of Chemistry, University of Illinois. It would have been impossible to have assembled this volume without their expert comments and advice. We also wish to thank Professor Joseph J. Lagowski for reading the volume and writing the introductory guide and overview (Chapter 1 of Part A) which brings such diverse subjects together. This volume would also have been impossible to complete without the efforts of Bonnie Koran, who produced most of the line drawings for the figures. We also acknowledge the help of many of our colleagues who have contributed helpful comments concerning this volume.

Cincinnati, Ohio
December, 1973

J. S. Mattson
H. C. Macdonald, Jr.
H. B. Mark, Jr.

CONTRIBUTORS TO THIS VOLUME

T. A. ATKINSON, Department of Chemistry, Michigan State University, East Lansing, Michigan

RONALD W. COLLINS, Department of Chemistry, Eastern Michigan University, Ypsilanti, Michigan

R. L. ELLIS, Department of Chemical Science, University of Illinois, Urbana, Illinois

K. JEFFREY JOHNSON, Department of Chemistry, University of Pittsburgh, Pittsburgh, Pennsylvania

J. J. LAGOWSKI, Department of Chemistry, The University of Texas, Austin, Texas

FREDERICK D. TABBUTT, The Evergreen State College, Olympia, Washington

MARTHA E. WILLIAMS, Coordinated Science Laboratory, University of Illinois, Urbana, Illinois

CONTENTS

CONTENTS OF PART B

COMPUTER-ASSISTED INSTRUCTION IN CHEMISTRY

(IN TWO PARTS)

Part A: General Approach

Chapter 1

THE USE OF COMPUTERS IN CHEMICAL EDUCATION
A Guide and Overview

J. J. Lagowski

Department of Chemistry
The University of Texas
Austin, Texas

I. INTRODUCTION

From the evidence presently available, the application of computer techniques may well have the most important impact on the educational process since the invention of the printing press. Computer-based methods of education and learning can potentially alleviate the logistics problems which characterize much of the present educational process. A discussion of these techniques appears in a subsequent section of this chapter. Basically, computers manipulate information; it is natural, therefore, that educators should consider seriously such devices in the educational process, which itself involves manipulation of information and the transfer of information to the student. Several important questions must be answered before computer techniques can be incorporated generally in a local environment. Are the techniques pedagogically viable and effective, i.e., do they do what they purport? How can useful programs developed in one institution be transferred to another? What is the cost of producing such programs, and what is the operating cost of a system dedicated to educational purposes? An attempt is made to answer these and other related questions in this volume.

II. COMPUTER-BASED TECHNIQUES

We can gain an appreciation of the ways in which the computer can be used to assist in the educational process by considering the nature of this process. Education involves teaching and learning. Often an instructor spends most of his time establishing learning situations for students rather than actually teaching them. Most of the computer-based techniques reported to date can be described as aids to (i.e., supplementary to) teaching and learning. Although there are several bases for classification of these methods, perhaps the most convenient is that based on the conventional division of work in the usual chemistry curriculum, *viz*. lecture, laboratory, recitation or discussion, homework problems, and examinations.

A. Tutorial/Drill Programs

In chemistry, as in all subjects, there are many pedagogical situations that require a patient tutor to guide the student through a logical sequence of steps in a (relatively) closed area with respect to information content. Usually the point is driven home by the use of numerous examples that are variations on a theme. Classically, such situations have been exploited by recitation or discussion periods and the use of homework problems. It is possible to write highly sophisticated programs for interactive computer systems which capture the detailed strategy that an individual teacher would use in a tutorial session. In essence that teacher (in the form of the program) is unbounded in space and time, since the program can be run at any time of day and can be transmitted to any geographic point the communications system permits. The alleviation of logistic constraints imposed by availability of rooms and/or instructors is obvious. Programs which may be classified in this category are discussed in detail in Chapters 1 and 6 of Part A and Chapter 1 of Part B.

B. Laboratory Simulation

Laboratory courses are usually designed to provide experience in experimental techniques as well as to develop a facility for manipulating raw data. Computer methods can provide experience in the latter phase of laboratory work but do little with respect to the former, unless the computer is treated as a laboratory tool interfaced between the student and the experiment. Other volumes in this series address the question of how the computer can be used in the laboratory to intervene in an experiment. Simulation programs are described in Chapters 2 and 3 of Part A and 1, 2, 5, and 6 of Part B. Basically, simulation programs can be designed to drive a student to make all the decisions he normally would make in a real experiment, i.e., collect data and draw conclusions based on the data. In essence, the driver program in a simulated experiment is a representation of a mathematical model for which the student sets parameters that have a manifestation in the real world. It is possible to introduce experimental errors in the results of computer-simulated experiments which reflect the accuracy expected from the equipment normally used in the laboratory as well as gross random errors. Thus, the students' experience in manipulating experimental results can be extended by simulating experiments that (a) might be conceptually simple but which require apparatus too complex for them to manipulate at a given educational level (e.g., atomic spectroscopy, x-ray diffraction), (b) consume a disproportionate amount of time if performed in their entirety, (c) are potentially dangerous, and/or (d) involve expensive equipment which cannot be supplied for large numbers of students on an individual basis. Simulated experiments can be used in various ways, e.g., as extensions of laboratory work as well as supplements to conventional lecture material (Chapters 1 and 6 of Part B).

It should be pointed out that the terminals on which students might perform simulated experiments need not be located in a laboratory; they may be placed in study rooms, library carrels, or even in dormitories. Further, simulated experiments need not be performed at the conventional laboratory periods, nor do the programs need to be available in an interactive mode. Success in using simulation programs in a batch environment has been demonstrated (Chapters 3 and 7 of Part B). Thus, a part of a student's laboratory experience can be provided in a more flexible manner than is now possible under the logistic constraints imposed by conventional laboratory work. The savings in costs for equipment and chemicals using computer-based methods is obvious.

C. Homework Problems

Homework problems have been used traditionally to provide the student with practice in problem-solving, but there are several difficulties with such strategy. Firstly, grading and record-keeping become important considerations in classes with large numbers of students. Secondly, feedback on problems (grading) is generally slow and individual error analysis nonexistent.

Many of the programs written for tutorial/drill mode or laboratory simulation contain drivers or subroutines well suited for generating problems. Separate homework-generating programs have also been developed (Chapter 8 of Part B). It should not be surprising that since programs can be written to generate homework problems, programs to grade homework are also possible, for, a program which can generate problems contains information on the details associated with common student errors. Thus, it is possible to give not only full credit for a problem, but also partial credit if intermediate steps are correct but the final answer is wrong. In addition, such grading programs usually incorporate subroutines which automatically yield appropriate statistical analysis of the student's performance on individual problems.

D. Examinations

An important element of computer-based techniques is the development of programs and data bases which generate examinations (Chapters 2 of Part A and 4 of Part B). The ability to generate a large number of examinations which are statistically equivalent gives the instructor a very powerful teaching device. With such programs available it is possible to create self-paced (Keller plan) courses for relatively large numbers of students; such flexible courses generally depend upon the availability of a "readiness test" on a student-demand basis. Input of the marked examinations into the grading program can be either in a batch mode or interactively using inexpensive mark-sense readers. The latter operation permits immediate feedback to the student. It is apparent that for large numbers of students at various points in a self-paced course the logistics of creating, administering, grading, and maintaining security for a large number of different examinations becomes virtually unmanageable using conventional methods. Computer-based methods not only expedite all aspects of examination-taking, but they can also incorporate relatively sophisticated record-keeping and statistical routines with little additional difficulty.

E. Other Uses of the Computer

The most obvious computer uses, e.g., as a retriever of information (Chapter 5 of Part A) or as a calculator, can also be included in the educational process. For example, we can now make available to students on a routine basis rather sophisticated data-analysis programs to assist them in organizing raw experimental data (Chapters 2 of Part A and 1, 3, 7 of Part B). In these situations, the computer essentially plays the role of a very sophisticated slide rule. Students can be given "canned programs" which perform detailed analysis, or they can be asked to write their own programs. The controversy about the pedagogical usefulness of allowing students to use canned programs rather than making them write their own programs has many of the characteristics of the arguments of several years past about the desirability of having students use a double-beam balance in undergraduate quantitative analysis laboratory when single-pan automatic balances were available. The arguments surrounding the use of "canned" programs will undoubtedly be resolved similarly.

F. Impact of Computer-Based Techniques on Education

The existence of the wide variety of computer-based techniques for teaching and learning outlined in this volume has a profound implication on the future of the educational process. The flexibility of providing a viable educational experience for students encompassing a wide spectrum of interests, abilities, and backgrounds is inherent in computer-based methods. It is now possible to create an educational system that is a continuum of experiences rather than a series of experiences quantized in time (semesters, terms, etc.). A reasonably large portion of a student's educational experience need not occur at a particular geographical location where historical accident dictated the establishment of an institution of learning. An educational system based on computer techniques can accept a student at any level and allow him to progress along any one of several (not necessarily specified) paths. Such a system would be useful for continuing education in this respect. In addition, the educational content of the system can be rapidly updated by changing individual programs, which is a difficult and time consuming process in conventional systems of education.

G. Transferability

Given that the arguments here are sufficiently compelling (at least in certain aspects) to encourage incorporation of some of these methods in a teaching situation, what is necessary to accomplish this task? Obviously, it would be desirable to take advantage of previous work in this area to the fullest extent. Transferability must be considered at several levels: technical, from the point of view of computing, and pedagogical, from the point of view of the environment (in the class, in the curriculum, by the student) for which the programs were successfully designed and used. The implication here is that the objectives of the program have been evaluated in one environment before transfer to another is attempted. The technical aspects of transferability revolve about the characteristics of the computer system on which the programs were developed compared with those of the system to which they are to be transferred. It is safe to say that presently there are very few computer systems which have completely compatible software; invariably, technical problems arise which are system-oriented when a program developed on one system is run on another. Thus, a spectrum of situations is possible, extending from one extreme where the systems involved are virtually the same and use a common language [e.g., IBM equipment using APL (Chapter 1 of Part B)], to the other extreme where equipment from different manufacturers running on different languages is involved. In the first instance the technical problems associated with transferability are at a minimum, but at the other extreme the problems are maximized. In any case it is possible to transfer the essence of the program in terms of a well documented flow diagram, so that local personnel can encode it using the available software. This, however, can be a time-consuming process. An attempt has been made to establish an organization* which will define the elements of transferability. In addition, an educational program repository for chemistry has been established,† and numerous educationally useful programs have been described in several meetings devoted to this subject [1-4]. There is, of course, the possibility of using programs in a local environment by teleprocessing from remote computers. This strategy eliminates the problems associated with recoding programs, but the communications costs can become unbearable.

Finally, the question of production costs, i.e., continued use by students for extended periods of time, must be faced. Chapter 5 of Part A addresses itself to some of these costs using a very large system dedicated to educational uses (Chapter 2 of Part B). Using this approach one would have to be prepared to make a significant capital outlay. On the other hand, there are data available suggesting that reasonable costs per student hour can be attained on a general purpose time-sharing system [5]. Information is also available [6] which suggests that time-sharing minicomputers with as little as 16K of core operating with extended basic are economically feasible main frames for many of the educational uses described here.

Finally, most of the discussion concerning interactive systems here has been in terms of use by individual students. It has been demonstrated (Chapters 4 of Part A and 6 of Part B) that computers can be successfully and meaningfully used as an aid in a lecture environment.

ACKNOWLEDGMENT

I should like to acknowledge the support of The Moody Foundation, The Esso Foundation, and Project C-BE, which is funded by the National Science Foundation, for providing generous support at various times over the past five years. Much of the information and many of the ideas expressed in this paper were developed during this period.

REFERENCES

1. "Proceedings on Conference on Computers in Chemical Education and Research," Northern Illinois University, De Kalb, Illinois, 1971.

---

*CONDUIT is an NSF sponsored project. Further information on chemistry programs in this project can be obtained from Dr. Joseph R. Denk, Curriculum Development Manager, CONDUIT, North Research Triangle Park, North Carolina 27709.

†Further information can be obtained from Professor R. Collins, Department of Chemistry, Eastern Michigan University.

1. Computers in Chemical Education

2. "Proceedings of the 1972 Conference on Computers in the Undergraduate Curricula," Southern Regional Education Board, Atlanta, Georgia, 1972.

3. "Proceedings of the 1972 Simulation Conference," June 14-16, 1972, Town and Country Hotel, San Diego, California, Simulation Council, Inc., La Jolla, California.

4. "Proceedings of the Second Conference on Computers in the Undergraduate Curricula," Computer Oriented Materials Products for Undergraduate Teaching (COMPUT), Dartmouth College, Hanover, New Hampshire, 1971.

5. S. J. Castleberry, G. H. Culp, and J. J. Lagowski, J. Chem. Ed., in press.

6. S. J. Castleberry, private communication.

Chapter 2

CURRICULUM ENRICHMENT WITH COMPUTERS

K. Jeffrey Johnson

Department of Chemistry
University of Pittsburgh
Pittsburgh, Pennsylvania 15260

I. INTRODUCTION

Since January, 1969, my research has involved implementing computing techniques in the chemistry curriculum. The projects under investigation include: computer-assisted instruction, CAI; the use of Fortran IV programs to simulate chemical systems and for data reduction; the use of an interactive graphics system; a computer-generated repeatable examination system; and an attempt to introduce computer programming as part of the chemistry curriculum at several levels.

The computer on which this work has been implemented is a PDP-10, dedicated to time-sharing. There are approximately thirty public terminals on the campus. There are three additional terminals in the chemistry building. The program languages used include: PIL, Pitt's Interpretive Language [1]; CATALYST/PIL, Pitt's CAI language [2]; Fortran IV; and PIGS, Pitt's Interactive Graphics System, a subset of the Culler-Fried interactive graphics system implemented at the University of California at Santa Barbara [3].

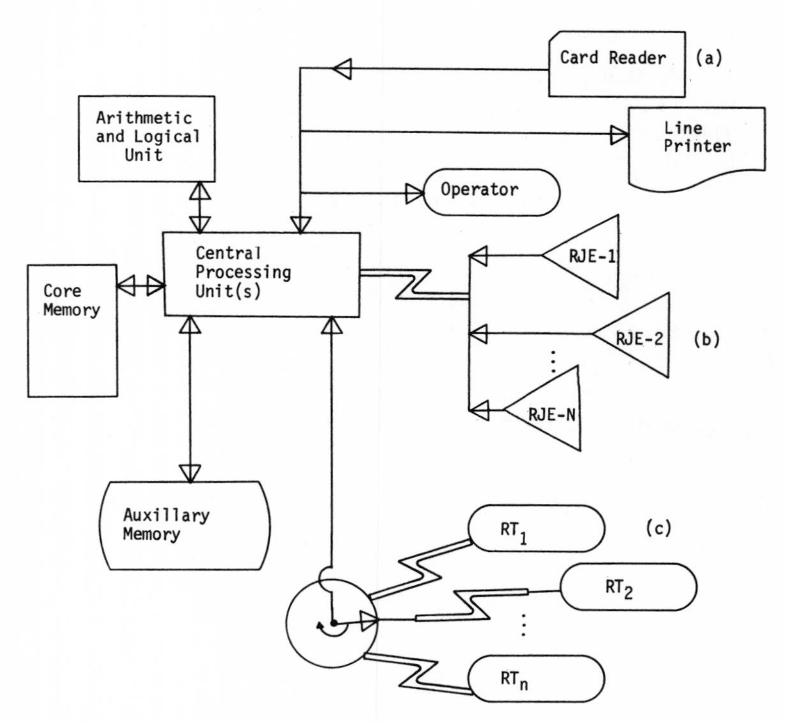

Fig. 1. Block diagram of a computer showing (a) batch processing, (b) remote-job-entry, and (c) time-sharing.

Figure 1 is a block diagram of a computer configured to support batch processing, remote-job-entry (RJE), and time-sharing. A time-sharing system discussed in this chapter could be implemented on an RJE or batch-processing system. Many of the programs can be implemented on a desktop programmable calculator.

The first section of this chapter describes the CAI project. Then a threefold approach to problem-solving with computers is described: the use of canned programs for data reduction and simulation, teaching of computer programming, and several applications of computer graphics. The third section is a preliminary report on an automated testing system. The chapter concludes with a discussion of the results to date and plans for the future.

II. COMPUTER-ASSISTED INSTRUCTION

A. Introduction

A number of chemistry CAI systems have been described in the literature [4]. The objective is to establish a dialogue between a student and a tutor, the latter being represented by the CAI program. This dialogue is frequently pitched at a remedial tutorial drill level; however, the degree of sophistication is the arbitrary decision of the lesson designer. The CAI program provides information, presents questions and problems, patiently awaits answers, gives immediate reinforcement, provides hints if help is requested, and branches to one of an arbitrary number of coded messages contingent upon the student's response.

CAI is particularly useful in large freshman chemistry classes. A significant number of students in these classes require outside assistance to master such fundamental concepts as stoichiometry and the gas laws. Frequently, it is a matter of sharpening dull arithmetic skills rather than clarifying chemical and physical concepts. The CAI system described here attempts to alleviate this problem. The students in our general chemistry sections are invited to access 26 CAI lessons that review stoichiometry, gas laws, colligative properties, Faraday's laws, solution stoichiometry, periodic properties, bonding, the Bohr atom, aqueous equilibrium, and kinetics. These lessons represent at least eight hours of (free) tutoring.

The programs are written in CATALYST/PIL, a lesson designer-oriented language. The students access the programs from any of the 30 public terminals on campus. The system is available nearly 24 hours a day, seven days a week.

The current (June, 1972) versions of these lessons have four features that make them particularly useful for tutorial drill applications in general chemistry. First, the student may type "HELP" at any point and receive a tutorial message. Because most of the terminals are teletypes rather than cathode-ray tubes, it is expensive in paper and time (terminal and computer) to print lengthy tutorial messages; furthermore, many students neither need nor want these detailed explanations. Therefore, such detail is provided only on demand. The following excerpt from a lesson on atomic and molecular weight calculations illustrates how the system works. (CATALYST uses the symbol ">" for the prompting character.)

```
    NOW LET'S CALCULATE THE ATOMIC WEIGHT OF A HYPOTHETICAL
    ELEMENT. THE ELEMENT HAS TWO ISOTOPES:

      ISOTOPE          EXACT MASS     % NATURAL ABUNDANCE

        43               42.9637              51.27
        45               44.9752              48.73

    WHAT IS THE ATOMIC WEIGHT OF THE ELEMENT?

>HELP

    NEED A HINT?

    THE ATOMIC WEIGHT OF AN ELEMENT CAN BE FOUND FROM THE
EXACT MASS AND THE ABUNDANCE OF THE ISOTOPES ACCORDING TO
THE EXPRESSION:

        (EXACT MASS)*(% ABUNDANCE/100) FOR ISOTOPE ONE
        +(EXACT MASS)*(% ABUNDANCE/100) FOR ISOTOPE TWO
        +...
        = ATOMIC WEIGHT OF THE ELEMENT

    NOW TRY IT!

>42.9637*.5127 + 44.9752*.4873

    VERY GOOD
```

This example also illustrates the second feature of the CATALYST/PIL system — the ease with which arithmetic problems can be handled. CATALYST will accept answers in the form of numerical expressions. For example,

```
    WHAT IS THE MOLAR VOLUME (IN LITERS) OF AN IDEAL GAS AT
    25.0 ATMOSPHERES PRESSURE AND 150 DEGREES C?

>(22.4*(150+273))/(25*273)

    RIGHT!
```

The PIL programming language is available for more involved computations. PIL is loaded with the students type "CALC". CATALYST is reloaded and execution continues from the appropriate place when "RETURN" is typed. Two examples follow. (PIL uses "*" for the prompting character.)

```
    WHICH OF THE FOLLOWING MINERALS IS RICHER IN COPPER?

        MINERAL                          MOLECULAR WEIGHT
        (CU)5FE(S)4                      502 GRAMS/MOLE
        (CU)2S                           159 GRAMS/MOLE

    (HINT: DETERMINE THE PERCENT CU IN BOTH MATERIALS, AW
    OF CU = 63.5)

>CALC

    READY:

*TYPE 5*63.5*100/502
5*63.5*100/502 = 63.24701
*TYPE 2*63.5*100/159
2*63.5*100/159 = 79.87421
*RETURN

    RESPOND TO LAST QUESTION

>CU2S

    CORRECT. NOW, ...
                  .
                  .
                  .
    WHAT IS THE (APPROXIMATE PH OF A 0.147 MOLAR ACETIC ACID
    SOLUTION? (KA = 1.8E-5)

>CALC

    READY:

*TYPE - THE LOG OF THE SQUARE ROOT OF (0.147*1.8E-5)
- THE LOG OF THE SQUARE ROOT OF (0.147*1.8E-5) = 2.788705
*RETURN

    RETURN TO LAST QUESTION

>2.79

    VERY GOOD
```

The desk-calculator mode of PIL is very easy to use. The students are given a three-page handout that describes the log-in/log-out procedure, the procedure for loading the CAI lessons, and several examples of PIL arithmetic expressions. This is an adequate introduction for the majority of the students. The ease with which chemical arithmetic problems can be solved is the most attractive feature of these lessons to many students.

The third feature of the CAI lessons is their extensive branching capability. In addition to "CALC" and "HELP" an arbitrary number of anticipated incorrect student responses can be coded and appropriately responsive messages printed. For example,

NOT QUITE. YOU'RE OFF BY A FACTOR OF 2. CHECK THE REACTION STOICHIOMETRY AND TRY IT AGAIN!

NOPE. YOU FORGOT TO CONVERT FROM DEGREES CENTIGRADE TO DEGREES KELVIN. HANG IN THERE!

YOU'RE GETTING WARM. YOU SHOULD HAVE USED THE ATOMIC WEIGHT OF NITROGEN INSTEAD OF THE MOLECULAR WEIGHT. TRY IT ONCE MORE.

This branching feature makes CAI an extremely powerful teaching aid. It is also possible to monitor student progress through the lessons. If most students miss a given question, it can be examined for ambiguities and edited as necessary. The effectiveness of a tutorial message can be easily checked by recording whether the student responded correctly after receiving the hint. A CAI lesson can thus be "tuned" until it is an effective teaching aid. Such a monitoring system is currently being designed.

The fourth feature is that the lesson provides the detailed solution to a problem if the student is unable to provide the correct answer after several attempts. For example,

WHICH OF THE FINAL VALUES WOULD YOU LIKE TO CALCULATE: P2, V2 or T2?

>V2.

OK, HERE'S YOUR PROBLEM:

A FIXED AMOUNT OF AN IDEAL GAS UNDERGOES THE FOLLOWING CHANGES:

INITIAL CONDITIONS

P1 = 1.417E+03 TORR   P2 = 2.395E+03 TORR
T1 = 2.976E+02 DEGK   T2 = 3.491E+02 DEGK
V1 = 2.509E+01 L

NOW, WHAT IS V2? (IN LITERS)

⋮ (three incorrect responses)

HERE'S THE SOLUTION:

THE COMBINED GAS LAW CAN BE STATED AS:

(P1*V1)/T1 = (P2*V2)/T2

SOLVING FOR V2 WE FIND?

V2 = V1*(P1/P2)*(T2/T1)

NOW THE CONDITIONS FOR OUR PROBLEM ARE:

INITIAL CONDITIONS	FINAL CONDITIONS
P1 = 1.417E + 03 TORR	P2 = 2.395E + 03 TORR
T1 = 2.976E + 02 DEGK	T2 = 3.491E + 02 DEGK
V1 = 2.509E + 01 LITERS	

SO V2 = 2.590E + 01*(1.417E + 03/2.395E+03)*(3.491E+02/2.976E+02)
= 1.741E + 01 LITERS

This feature is particularly useful as a study aid. Pains are taken to provide the answer in enough detail so the student can grasp the solution either at the terminal or in his study.

Many students find these features very clever and are "turned on" to CAI immediately. Given adequate hardware and software support, it is clear that CAI can be of tremendous value to chemical educators. The rest of this section describes the CATALYST/PIL system and the CAI lessons in more detail. The section concludes with a discussion of the effectiveness of CAI.

B. CATALYST/PIL

CATALYST, Computer Assisted Teaching and Learning System, is a text-oriented processor designed for CAI lesson-writers who are not computer programmers. PIL, a JOSS-derived interpretive language, is very easy to learn and use. The CATALYST/PIL interface is a nearly ideal CAI language.

The lessons described here consist of several questions or problems organized linearly. That is, the student starts with the easiest question; if he answers this correctly, he proceeds to harder questions. If he fails to answer correctly in several attempts, the solution is provided and he is then branched to the next question. A simplified flow chart for a single question or problem is given in Fig. 2. In summary, the question is printed at the

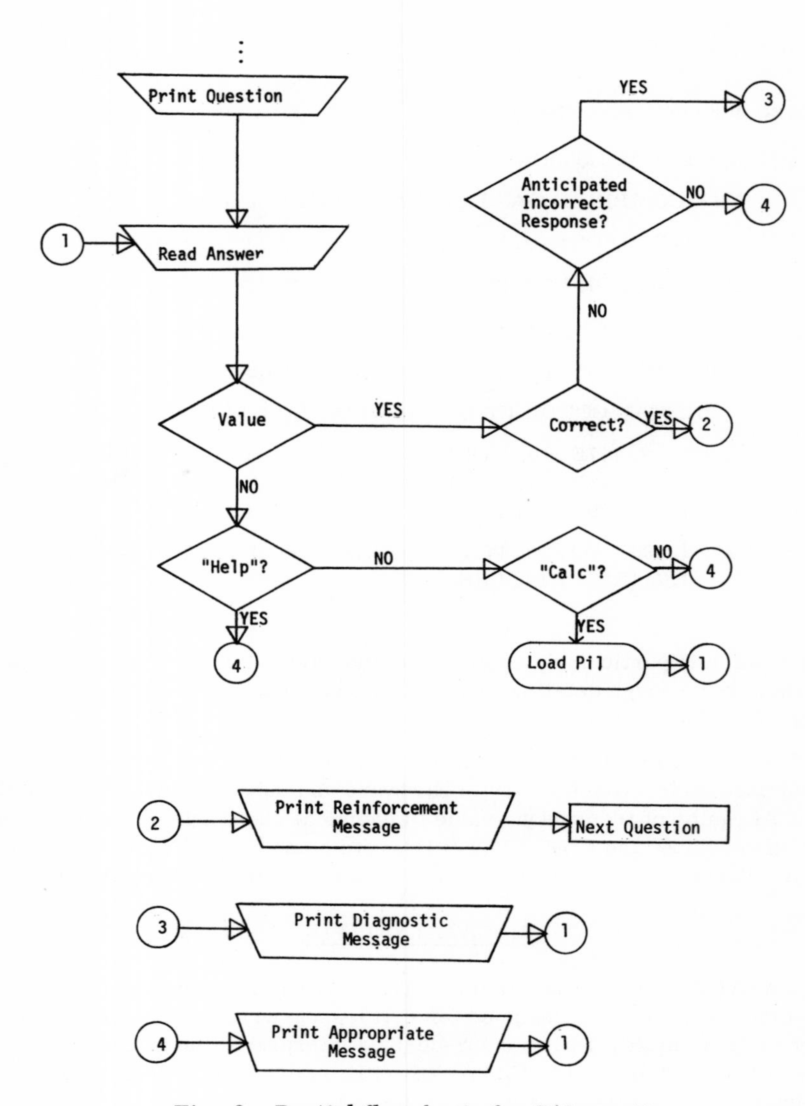

Fig. 2. Partial flowchart of a CAI question.

terminal and the student's response is read. PIL is loaded if the student types "CALC," a tutorial message is printed if he asks for "HELP," and the appropriate branch is made if his answer is correct, is one of the anticipated incorrect responses, or is some other answer. The student is given several attempts to answer correctly, but if he is unable to do so, then the program prints the solution for him.

The CAI lessons contain a mixture of CATALYST and PIL statements. For example,

```
        .
        .
        .
*FRAME 3
.SET C(4) = 1
    WHICH OF THE FOLLOWING ACIDS DISSOCIATES TO THE GREATEST
    EXTENT IN WATER? TYPE THE NAME OF THE ACID.

          CARBONIC ACID      (H2CO3)         K1 = 4.6E-7
          ACETIC ACID        (H4C2O2)        K = 1.8E-5
          BORIC ACID         (H3BO4)         K1 = 5.8E-10
          OXALIC ACID        (H2C2O4)        K1 = 6.2E-2

*FRAME 423
??FRAME 40 TYPE 2 HELP
?FRAME 46 TYPE 2 BORIC
?FRAME 100 RETURN 4 TYPE 2 OXALIC
    SORRY, THAT'S NOT THE ANSWER I'M LOOKING FOR. TRY AGAIN
    OR ASK FOR HELP.
.SET C(4) = C(4) + 1
?FRAME 41
*FRAME 46
.SET C (4) = C(4) + 1
    NO, YOU HAVE CHOSEN THE ACID WHICH DISSOCIATES THE LEAST.
    ARE THE EXPONENTS TROUBLING YOU?

                    100 > 10 > 0.1 > 0.0001
                    1.0E2 > 1.0E1 > 1.0E-4

    REMEMBER: THE GREATER THE DISSOCIATION CONSTANT, THE
    GREATER THE NUMBER OF IONS IN SOLUTION.

    TRY IT AGAIN.
?FRAME 423
*FRAME 501
        .
        .
        .
```

Frame 3 contains the text and commands between the statements *Frame 3 and *Frame 423. A period in column 1 denotes a PIL statement. Text that appears without a period, asterisk, or question mark in column 1 is printed character for character at the terminal. The control characters "??"

indicate that an answer is to be read from the terminal. Control is branched to Frame 40 if the answer contains the characters "HELP", 46 if the answer contains the characters "BORIC," and to Frame 100 if the answer contains "OXALIC." The statement "?FRAME 41" is an unconditional branch to Frame 41.

Most of the emphasis here is on numerical work. A more typical example is the following:

HOW MANY GRAMS OF NH4NO3 MUST BE ADDED TO 2.50 LITERS OF H2O TO PREPARE A 0.175 MOLAR NH4NO3 SOLUTION? (MW OF NH4NO3 = 80.0 GRAMS/MOLE).

The CATALYST statement,

```
??FRAME 600 TYPE 5 (0.99*ANS1) < RESPONSE & RESPONSE
                                 < (1.01*ANS1)
```

causes control to branch to Frame 600 if the student answers with the value stored in ANS1 (80*0.175*2.5) within a 1% tolerance. The PIL random number generator is used to provide randomization wherever appropriate. A student may try a given lesson several times and be given a different set of problems and questions each time.

C. The CAI Lessons

The 26 CAI lessons currently available are described in Table 1. They represent at least eight hours of (free) tutoring available to interested students. It can be seen from the description of the lessons that the emphasis is on chemical arithmetic. Stoichiometry and the gas laws are reviewed rather rapidly during the first few weeks of the first term of general chemistry. The relevant lessons (STOICH.1-4, EXPERT.1, COLLIG.1, ELECT.1-3, CONC.1-2, and GAS.1-3) are designed to help students review their high school chemistry. Students have commented that these lessons are very helpful in preparing for hour exams and in reviewing for the final exam. The remainder of the first term course is devoted to atomic structure, bonding, periodic trends, and descriptive chemistry. The lessons BOND.1-3 provide a drill on some of these topics.

The second term course usually covers thermodynamics, gas phase and aqueous equilibrium, kinetics, electrochemistry, more descriptive chemistry, and other topics as time allows. The seven lessons PH.1 and AQEQ.1-6 provide a number of questions and problems designed to help

TABLE 1

The CAI Lessons

Name	Description
STOICH. 1-4	Twenty problems reviewing molecular weight determinations, gram-mole and molecule-mole conversions, empirical formulas, percent composition, atomic weight of an element from its isotopic composition and weight-weight stoichiometry
EXPERT. 1	Simulation of a stoichiometry experiment to determine the empirical formula of a copper and sulfur compound
COLLIG. 1	Three problems drilling freezing point depression, boiling point elevation, and molecular weight determinations
ELECT. 1-2	Fifteen problems in electrochemistry including: mole-faraday, amp-coulomb, and current-mass conversions; electrolysis calculations; equivalent — electrical work conversions; cell potentials; and Nernst equation problems
GAS. 1-3	Twelve problems reviewing the gas laws (Boyle, Charles, Avogadro, Guy-Lussac, Graham, Dalton, and the combined gas law)
CONC. 1-2	Concentration units (molarity and molality)
PH. 1	pH - $[H^+]$ drill program
AQEQ. 1-6	Twenty questions and problems drilling aqueous equilibrium including properties of acids and bases, weak and strong electrolytes, dissociation equilibria, hydrolysis equilibria, pH, buffers, polyprotic acid equilibria, and solubility calculations
TITR. 1-2	Simulation of HCl-NaOH and CH_3COOH-NaOH titrations
KIN. 1	Simulation of the $A \underset{k_2}{\overset{k_1}{\rightleftarrows}} B$ system
BOND. 1-3	Twenty-one questions and problems reviewing periodic trends, quantum numbers, magnetic properties, and Bohr atom calculations

students master aqueous equilibrium. TITR.1 and 2 are simulation-drill lessons that help students calculate acid-base titration curves. KIN.1 leads students to the "discovery" of the equilibrium condition in the $A \underset{k_2}{\overset{k_1}{\rightleftarrows}} B$ system,

$$K_{eq} = \frac{[B]_{eq}}{[A]_{eq}} = \frac{k_1}{k_2} .$$

The lessons provide help upon request at every question, allow transfer to PIL for calculations when desired, utilize the multiple branching scheme discussed, and provide detailed solutions to the problems if the student fails to enter the correct answer after several attempts.

D. Discussion

Most of these lessons have been made available to freshman chemistry students for two years. There is no coercion — the students are free to take advantage of the lessons if they wish. The procedure is to acquire the CAI memo and a computer number from the stockroom attendant, read the memo carefully, find a free computer terminal, log on, load a CATALYST lesson, execute it, and log off. There are programming assistants available at the thirty public terminal sites if the students need them. Also, I invite them to call or see me at any time should they have a problem with the CAI lessons. Of 600 students in two sections of the first-term general chemistry course, approximately one-third successfully executed at least one CAI lesson during the Fall 1970 term and approximately one-half did so during the Fall 1971 term. Part of this increase is due to the greater number of public terminals available (15 in 1970, 30 in 1971), but part is because the word is getting out that these CAI lessons are both fun and effective. I fully expect the usage to continue to increase.

Plans to significantly improve the lessons are being implemented. They are being restructured according to the flowchart given in Fig. 3. The students will first be presented with a question of intermediate difficulty (0) and then branched to more difficult (+1, +2, etc.) or less difficult (-1, -2, etc.) questions contingent upon his responses. Other features, including "SKIP," "BACK," and "COMMENT," are being coded into the CAI lessons for more versatility. If a student wishes to skip the question just presented, he will type "SKIP" or "+1" and be branched to the next question. The "BACK" feature will allow a student to attempt an easier problem. The "COMMENT" feature will enable a student to type his opinions and suggestions at the terminal.

Also, a complete management system is being developed to provide feedback on lesson use. Special PIL datasets will be loaded when execution of a CATALYST lesson begins. These PIL datasets will record which questions are answered correctly or incorrectly how often a help or hint is used,

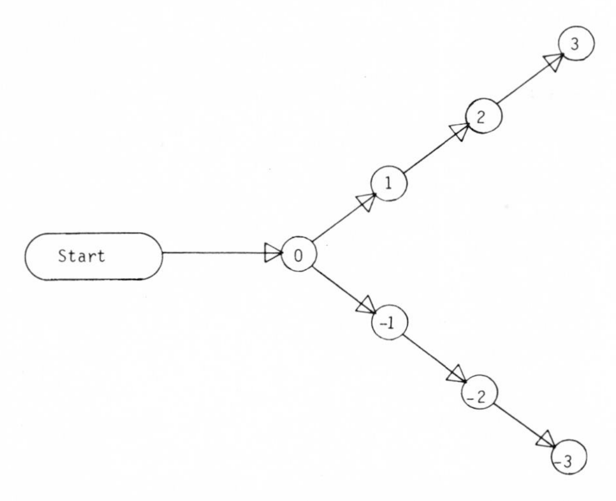

Fig. 3. Flow of a CAI lesson with nonlinear branching.

and how effective it is (i.e., whether the question is answered correctly or incorrectly after the hint is presented), normal and abnormal terminations of the lessons, the total number of right and wrong responses, the number of times each question is skipped or repeated, and the comments and suggestions typed by the student. This information will pinpoint problem areas in the lessons and help us create more effective teaching modules.

So far student feedback has been encouraging. The students enjoyed using the lessons and are asking for more programs. They have indicated on attitude questionnaires that the time required to use the programs was well spent and that CAI is an effective teaching tool. They also recommend continued development of CAI lessons in freshman chemistry and indicate that they would participate in similar programs in other courses if given the opportunity. The following student comments are taken from CAI questionnaires:

"I found working with the computer very interesting and enjoyable. Because of this experience, I am thinking of taking some courses in computer programming."

"The idea of computer-assisted (instruction) program is great ..."

"Once you've read the material, and understand it (outside of the computer room), I feel it can be an invaluable study aid.... Thank you for the opportunity."

"It gives me more confidence in my work and motivates one more. I find that I'm enjoying chemistry a lot more because of it."

"I sincerely hope that the computer program will be continued and expanded for the rest of Chem. 11 and onto Chem. 12 material."

"... I am really glad that ... CATALYST was offered to me and I hope that more CATALYST will be offered for the rest of Chem. 11 and Chem. 12."

"I have also found this method very effective in my biology course."

"The computer is very witty!"

Predictably, the students most enthusiastic about CAI are the upper 20% of the class. Yet the emphasis is on tutorial/drill lessons for the average and below-average students. Most of the students in this latter group "can never find the time" to locate a terminal and try out the programs. I hope that with increased availability of terminals and with the development of a more complete and effective library of CAI programs, this motivation problem will be solved. My group will continue to develop CAI lessons. We think that with care we can provide our freshman chemistry students with a learning aid that will make the study of chemistry more interesting and enjoyable.

III. COMPUTER-AUGMENTED LEARNING

A. Introduction

The computer-assisted instruction lessons described in the preceding section provide a tutorial/drill facility for students who require remedial instruction in chemistry. In this section, several approaches to computer-augmented learning, CAL, will be described. The distinction between CAI and CAL is roughly analogous to dual versus solo mode flight instruction [5]. In dual mode the student interacts with his instructor who directs the learning process. In solo mode the student is at the controls. Simulation and data reduction programs are therefore classified as dual mode, even though the objectives are more subtle than drill and practice CAI. The student who writes a program or uses a canned program as a tool to solve a chemistry problem is performing in solo mode.

As the student progresses through the Pitt chemistry curriculum, he is given several opportunities to use computers in both dual and solo modes. In addition to CAI, freshmen (and all other interested chemistry students) are invited to attend an optional, no-credit evening short-course in PIL programming. This experience enables the students to solve some interesting and stimulating optional assignments in their courses. Also, several simulation programs are made available to students in analytical and physical chemistry. A computer applications course is available for upper-division students, and there is an interactive graphics terminal for more sophisticated simulation work. These approaches to curriculum enrichment with computers will now be described in more detail.

B. The Optional PIL Course

This no-credit, no-grade course is an invitation to all well-motivated students (freshman through faculty) who want to learn something about computers and how they can be used to solve chemistry problems. It is offered every term, one night a week for four weeks. Several PIL programs are distributed as handouts. The fundamentals of computer programming are discussed: input and output, looping, testing, branching, etc. The desk calculator mode of PIL is briefly reviewed, and students are invited to use PIL to solve their homework problems. The emphasis, however, is on the stored-program mode. The following example is a PIL program to calculate the compressibility factor of a real gas at its critical temperature using the van der Waals equation of state. For one mole,

$$\left(P + \frac{a}{V^2}\right)(V - b) = RT,$$

where P is the pressure in atmospheres, V is the volume in liters, T is the temperature in degrees Kelvin, R is the gas constant (0.0821), and a and b are the van der Waals parameters for the real gas. This equation is linear in P but cubic in V. Therefore, calculations of the compressibility factor (Z = PV/RT) are facilitated by computing P over a defined range of V:

$$P = \frac{RT}{V - b} - \frac{a}{V^2}, \quad V_o \leq V \leq V_f.$$

The program reads a, b, V_o, V_f, and V_{inc}, the desired volume increment. The critical temperature is 8a/27Rb. The volume is initially set to V_o and then incremented by V_{inc} until V_f is exceeded. The sample execution is for CO_2, $0.05 \leq V \leq 1$ liter.

```
10.1      * PROGRAM TO GENERATE A TABLE CONTAINING
10.15     * VOLUME, PRESSURE AND COMPRESSIBILITY OF A
10.2      * VAN DER WAALS GAS
10.25     *
10.3      DEMAND A,B,VO,VF,VINC
10.35     TYPE A,B,VO,VF,VINC
10.4      SET R = 0.0821, T = 8*A/(27*R*B), RT = R*T
10.45     TYPE T
10.5      SET FMT ="    ..........      ..........      .........."
```

```
10.55    FOR I = 1 TO 2:LINE
10.58    TYPE"      VOLUME        PRESSURE       Z = PV/RT"
10.6     LINE
10.69    FOR V = VO TO VF BY VINC:  DO PART 20
10.98    LINE
20.1     SET P = RT/(V - B)-A/V/V, Z = P*V/RT
20.2     TYPE IN FORM FMT,V,P,Z
```

```
*DO PART 10
A = >3.59
B = >.0427
VO = >.05
VF = >1
VINC = >.1
Z = 3.59
B = 0.0427
VO = 0.05
VF = 1.0
VINC = 0.1
T = 303.4238
```

VOLUME	PRESSURE	Z = PV/RT
5.000E-02	1.976E+03	3.967E+00
1.500E-02	7.260E+01	4.371E-01
2.500E-01	6.272E+01	6.295E-01
3.500E-01	5.175E+01	7.272E-01
4.500E-01	4.343E+01	7.845E-01
5.500E-01	3.723E+01	8.221E-01
6.500E-01	3.252E+01	8.485E-01
7.500E-01	2.883E+01	8.682E-01
8.500E-01	2.588E+01	8.833E-01
9.500E-01	2.347E+01	8.953E-01

This program can generate the data for compressibility-pressure plots like the ones plotted in Fig. 4. Inquisitive students can use this program to explore the van der Waals model and painlessly discover how sensitive the compressibility function is to changes in a, b, and T ($T \geq T_c$). One student who took this optional evening short course wrote a program to find the pressure ($P > 0$) for which Z = 1.00. He then wrote a program to compute the area under the compressibility function and attempted to correlate that area with a, b, and T. The students are invited to explore other equations of state [6].

There are several interesting chemical systems that can be simulated using very simple PIL programs. For example, the relationship between

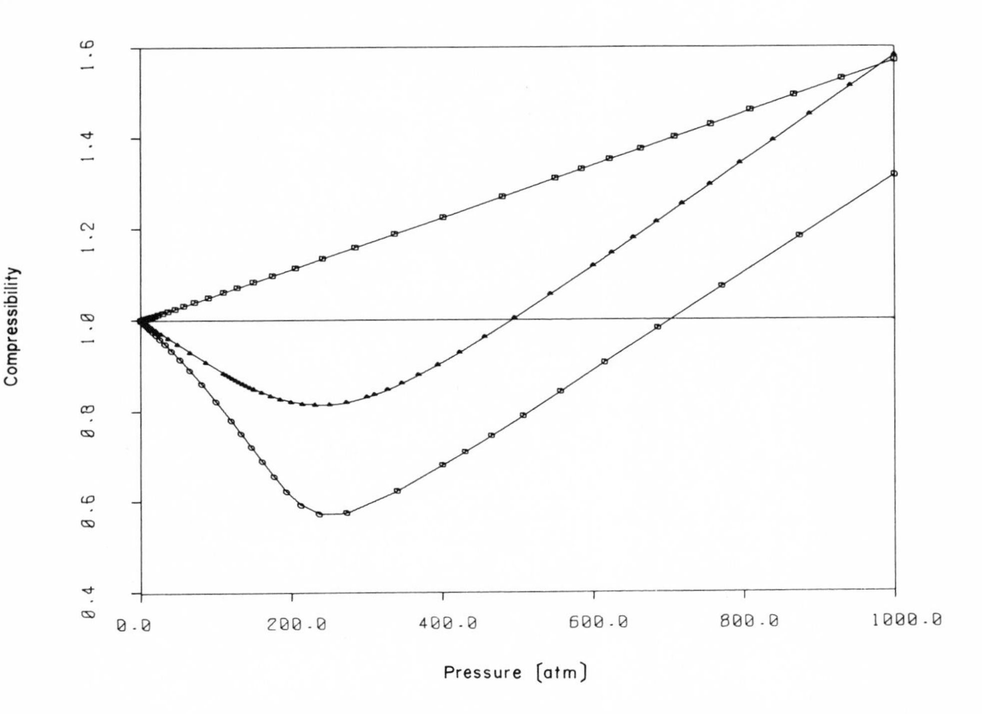

Fig. 4. Compressibility factor of He (top), CO_2, and NH_3 (bottom) at 500°K.

the rate constants and the equilibrium constant for the system

$$A \rightleftarrows B \qquad K_{eq} = \frac{[B]_{eq}}{[A]_{eq}} = \frac{k_1}{k_2}$$

can be easily "discovered" by computing concentration-time profiles using the equations

$$A(t) = A_0 - m[1 - \exp\{-(k_1 + k_2)t\}],$$

$$B(t) = B_0 + m[1 - \exp\{-(k_1 + k_2)t\}] = A_0 + B_0 - A(t),$$

where A_0 and B_0 are initial concentrations and $m = (k_1A_0 - k_2B_0)/(k_1 + k_2)$. The output of such a program is illustrated graphically in Fig. 5. Other examples include: solubility of salts as a function of pH and complexing ligands, acid-base equilibrium problems, overlapping radiodecay curves, Bohr atom calculations, etc.

This optional evening short course attracts many of the more highly motivated students. It provides them with the ability to penetrate some

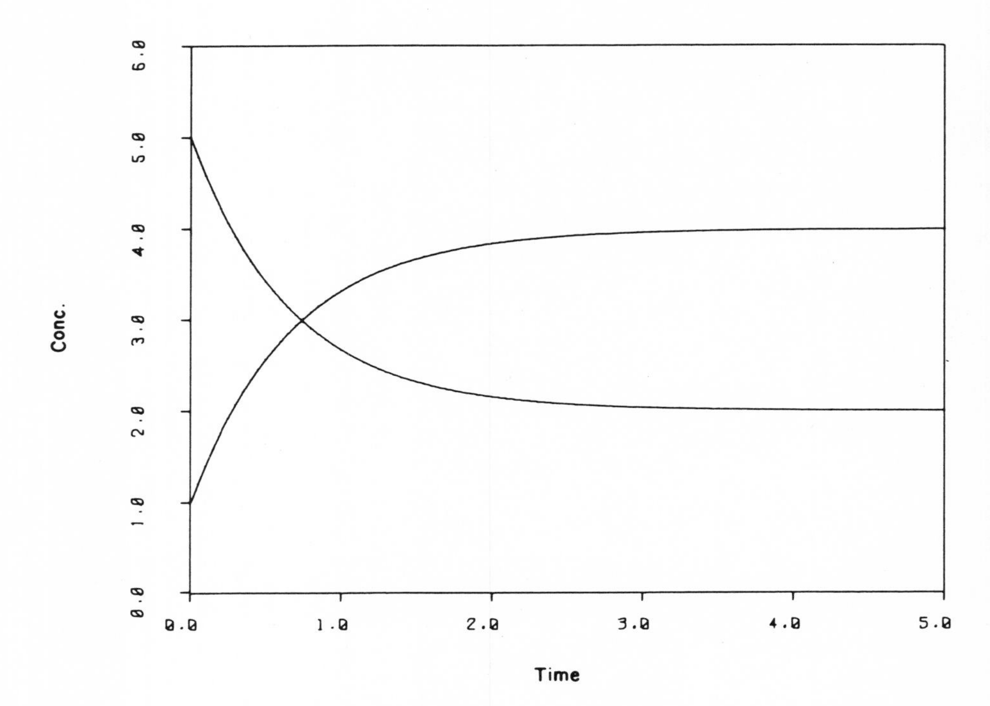

Fig. 5. Concentration-time plot for the $A \underset{k_2}{\overset{k_1}{\rightleftarrows}} B$ system. ($k_1 = 1$, $k_2 = 0.5$, $A_0 = 5.0$, $B_0 = 1.0$, A_∞ 2.0, $B_\infty = 4.0$).

chemical concepts more deeply and to take on problems that go beyond the exercises in their textbooks.

C. Simulation and Data Reduction Programs

The easiest way to introduce computing techniques in the chemistry curriculum is with canned simulation and data reduction programs. The Journal of Chemical Education and the proceedings from several conferences on the uses of computers in education give examples of computer programs used for simulations and for laboratory data reduction [7]. The programs may be used as black boxes, so there is no need for special languages and hardware. Many programs of this type have been implemented on desktop programmable calculators.

The data reduction programs allow a more penetrating analysis of data than students achieve working only with slide rules and calculators. For a trivial example, a linear regression analysis program can provide not only the intercept and slope of the straight line that gives the best fit to the data in the least squares sense, but it can also provide the standard deviations of the intercept and slope, the correlation coefficient and a table containing the observed data, the calculated ordinates and the difference between observed and calculated ordinates. A rate constant can therefore be reported as $1.0 \pm .2 \times 10^{-2}$ sec^{-1} instead of 1.027×10^{-2} sec^{-1}.

Simulation programs allow students to explore a mathematical model that approximates a chemical system. The model can frequently be represented by a function,

$$y = f(x; \alpha_1, \alpha_2, \ldots, \alpha_p),$$

where y and x are the observable dependent and independent variables, and the alphas represent adjustable parameters. For example, the M^{2+}-EDTA titration system is a three-parameter problem,

$$pM = f(V; K_f, pH, C_m),$$

where pM is the negative logarithm of the free metal ion concentration, V is the volume of titrant EDTA, K_f is the formation constant of the M^{2+}-$EDTA^{4-}$ complex, pH is the pH at which the titration is buffered, and C_m is the initial concentration of the metal ion. One use of an EDTA titration simulation program would be to let students verify hand-calculated values as part of a homework assignment. A considerably more exciting application is that simulation programs allow the more inquisitive students to painlessly get answers to the question, "what if?". A program that simulates EDTA titrations will allow an interested student to discover how the shape of the titration curve varies with systematic changes in K_f, pH, and C_m.

Table 2 contains a brief description of four simulation programs and one data reduction routine that have been used in the sophomore-level quantitative analysis course at Pitt. These programs are written in Fortran IV. Core images are saved on the disk. Those students who are interested in using these programs are provided with a computer number and the information required to execute the programs.

The program SOL calculates the solubility of the sparingly soluble salt, M_zY_x, as a function of pH:

$$M_zY_{x(s)} \rightleftarrows z'M^{x+} + x'Y^{z-}.$$

Here the anion, Y^{z-}, hydrolyzes to form z-conjugate acids:

$$Y^{z-} + H_2O \rightleftarrows HY^{(1+z)-} + OH^-,$$

$$HY^{(1+z)-} + H_2O \rightleftarrows H_2Y^{(2+z)-} + OH^-$$
$$\vdots$$
$$H_{z-1}Y^- + H_2O \rightleftarrows H_zY + OH^-.$$

TABLE 2

Analytical Chemistry Programs

Name	Description
SOL	Calculate the solubility of M_zY_x as a function of pH. The anion, Y^{z-}, hydrolyzes to form z acids, $HY^{(1+z)-}$, $H_2Y^{(2+z)-}$, $\cdots$ H_zY. Parameters: z, x, K_{sp}, K_1, K_2, $\cdots$ K_z
HA	Simulate the weak acid-strong base titration system. Parameters: K_a, C_a.
EDTA	Simulate the M^{z+}-EDTA titration system. Parameters: K_f, pH, C_m.
NERNST	Calculate the EMF of the reversible electrode $M\|M^{z+}$ in an aqueous ammonia solution as a function of pH. Parameters: z; the standard reduction potential for $M^{z+} + ze^- \rightleftarrows M$; n, the number of ammine complexes formed; and K_i, i = 1, 2, ... n—the stepwise formation constants.
TITR	Data reduction program that determines the thermodynamic acid dissociation constants for β-alanine from the student's titration data by making the appropriate activity corrections.

The stoichiometric coefficients z' and x' are not necessarily equal to z and x. For example,

$$CaC_2O_{4(s)} \rightleftarrows Ca^{2+} + C_2O_4^{2-} \quad (x = z = 2,\ x' = z' = 1),$$

$$Ag_3AsO_{4\,(s)} \rightleftarrows 3Ag^+ + AsO_4^{3-} \quad (x = x' = 1,\ z = z' = 3).$$

The solubility of Ag_3AsO_4 as a function of pH is

$$S = \left[\frac{K_{sp}}{27} \left\{ 1 + \frac{[H^+]}{K_3} + \frac{[H^+]^2}{K_2 K_3} + \frac{[H^+]^3}{K_1 K_2 K_3} \right\} \right]^{1/4},$$

where S is the solubility, K_{sp} is the solubility product for Ag_3AsO_4, and K_1, K_2, and K_3 are the stepwise dissociation constants for arsenic acid. Figure 6 shows plotted output from SOL for AgCN, Ag_2CO_3, and Ag_3AsO_4.

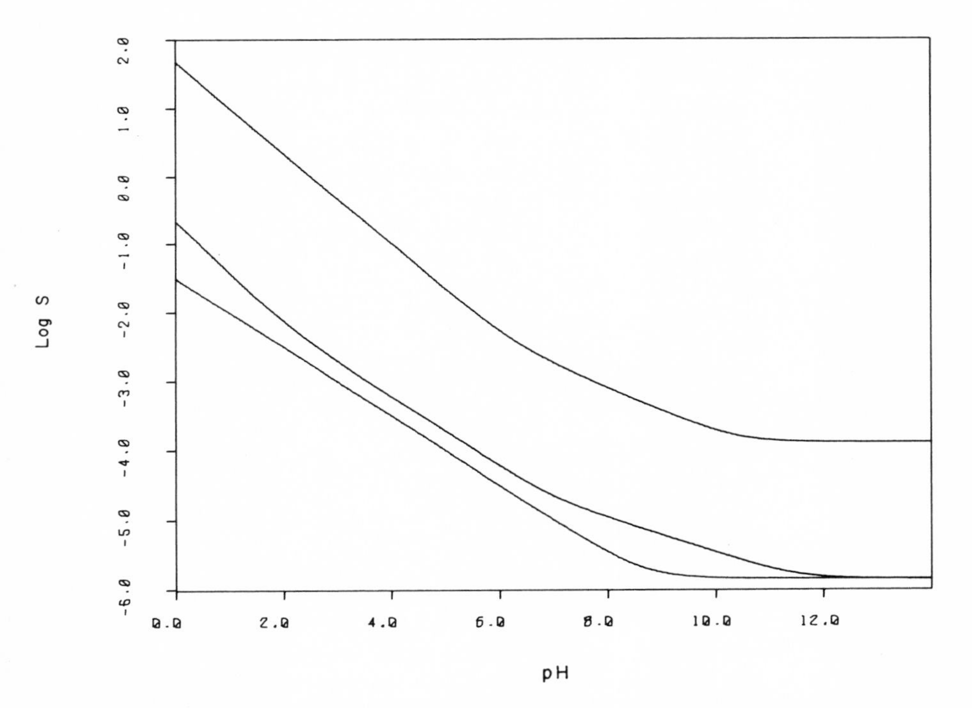

Fig. 6. Solubility of Ag_2CO_3 (top), Ag_3AsO_4, and AgCN (bottom) as a function of pH.

The program HA simulates the weak acid/strong base titration system. There are two adjustable parameters: K_a, the dissociation constant of the weak acid, and C_a, the initial weak acid concentration. The program also requires the student to enter the molarity of the strong base (titrant) and the volume of the acid solution. A sample execution of the program follows.

.RUN HA

ENTER MOLARITY OF ACID, MOLARITY OF TITRATING BASE, VOLUME OF ACID AND PKA OF ACID (4F) (ZERO ACID MOLARITY ENDS PROGRAM)

>.1, .1, 30, 5

CONCENTRATION OF ACID = 1.0000E-01
CONCENTRATION OF BASE = 1.0000E-01
VOLUME OF ACID = 3.0000E+01
PKA = 5.0000E+00

% TITR	PH	VOL OF BASE
0.00000E-01	3.002E+00	0.000E-01
2.03684E+00	3.399E+00	6.111E-01
5.70906E+00	3.796E+00	1.713E+00
1.33987E+01	4.192E+00	4.020E+00
2.79267E+01	4.589E+00	8.378E+00
4.91587E+01	4.986E+00	1.475E+01
7.06832E+01	5.382E+00	2.120E+01
8.57364E+01	5.779E+00	2.572E+01
9.37440E+01	6.176E+00	2.812E+01
9.73928E+01	6.572E+00	2.922E+01
9.89377E+01	6.969E+00	2.968E+01
9.95715E+01	7.366E+00	2.987E+01
9.98286E+01	7.763E+00	2.995E+01
9.99336E+01	8.159E+00	2.998E+01
9.99794E+01	8.556E+00	2.999E+01
1.00007E+02	8.953E+00	3.000E+01
1.00040E+02	9.349E+00	3.001E+01
1.00110E+02	9.746E+00	3.003E+01
1.00277E+02	1.014E+01	3.008E+01
1.00695E+02	1.054E+01	3.021E+01
1.01741E+02	1.984E+01	3.052E+01
1.04398E+02	1.133E+01	3.132E+01
1.11336E+02	1.173E+01	3.340E+01
1.30871E+02	1.213E+01	3.926E+01
2.00000E+02	1.252E+01	6.000E+01

ENTER MOLARITY OF ACID, MOLARITY OF TITRATING BASE, VOLUME OF ACID AND PKA OF ACID (4F) (ZERO ACID MOLARITY ENDS PROGRAM)

>0

EXECUTION TIME: 0.44 SEC.

Here the period is the PDP-10 time-sharing system prompting character. The program cycles until a value of zero for C_a is given. The PDP-10 Fortran free-format feature (4F) allows the student to enter the required data without being concerned with counting columns at the terminal. The program uses the Newton-Rapheson method [8] to solve the exact (cubic) equation to give the pH at 0% and 200% titration. This defines the pH range for the titration. The remaining points on the titration curve are more efficiently computed using the inverse function,

$$V_b = -V_a \left\{ \frac{[H^+]^3 + K_a[H^+]^2 - (K_w + K_aC_a)[H^+] - K_wK_a}{[H^+]^3 + (K_a + C_b)[H^+]^2 + (K_aC_b - K_w)[H^+] - K_aK_w} \right\}.$$

This method not only substantially reduces the execution time, but also provides most of the output in the region of the equivalence point. Figures 7 and 8 show the effect of systematic variation of C_a and K_a.

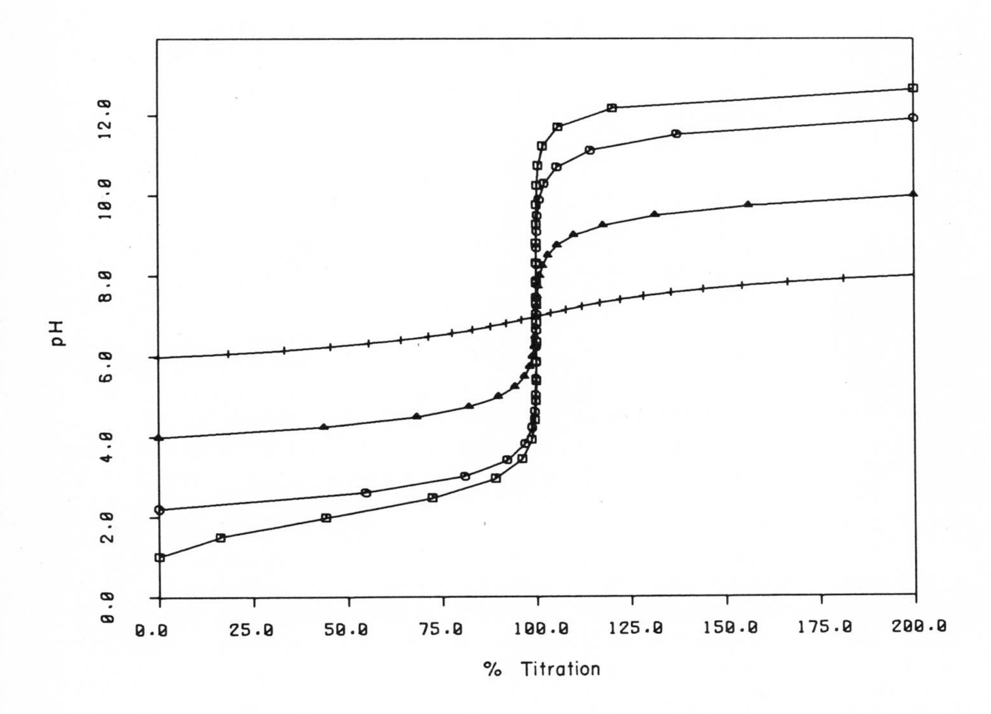

Fig. 7. Weak acid-strong base titration curves as a function of weak acid concentration. For $pK_a = 2$, C_a has the following values: 1.0 M (□), 0.01 M (○), 1.0×10^{-4} M (▲), and 1.0×10^{-6} M (+).

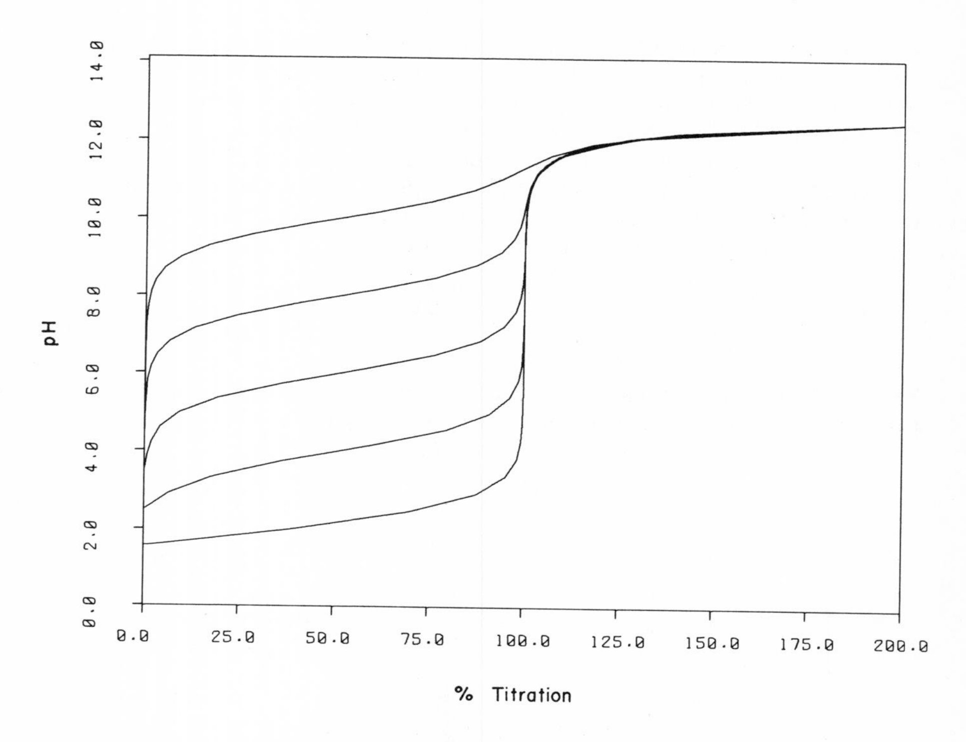

Fig. 8. Weak acid-strong base titration curves as a function of pK_a; $C_a = 0.1$ M, $pK_2 = 2$ (bottom), 4, 6, 8, and 10 (top).

The EDTA program simulates the M^{Z+} -EDTA titration system [9]. There are three parameters: the pH at which the system is buffered, the formation constant (K_f) of the M^{Z+}-EDTA complex, and the initial concentration of the metal ion. The program also asks for the concentration of EDTA titrant and the initial volume of the metal ion solution. Figures 9 through 12 were generated from values obtained from the EDTA program. Figure 9 demonstrates the effect of pH on the titration curve of a metal, initially 0.01 M, that forms an EDTA complex with formation constant 10^{15}. The four curves were calculated using pH = 5 (bottom), 7, 9, and 11 (top). Figure 10 demonstrates the effect of changing the M^{Z+}-$EDTA^{4-}$ formation constant. The titrations are buffered at pH = 11, and each metal is initially 0.01 M. The formation constants used are 10^8 (bottom curve), 10^{11}, 10^{14}, and 10^{17} (top curve). Figure 11 shows the effect of the initial metal ion concentration. The titration is buffered at pH = 10, and the formation constant is 10^{15}. The four curves are for initial metal ion concentration 10^{-2} M (bottom), 10^{-4}, 10^{-6}, and 10^{-8} (top). Figure 12 is a titration feasibility plot.

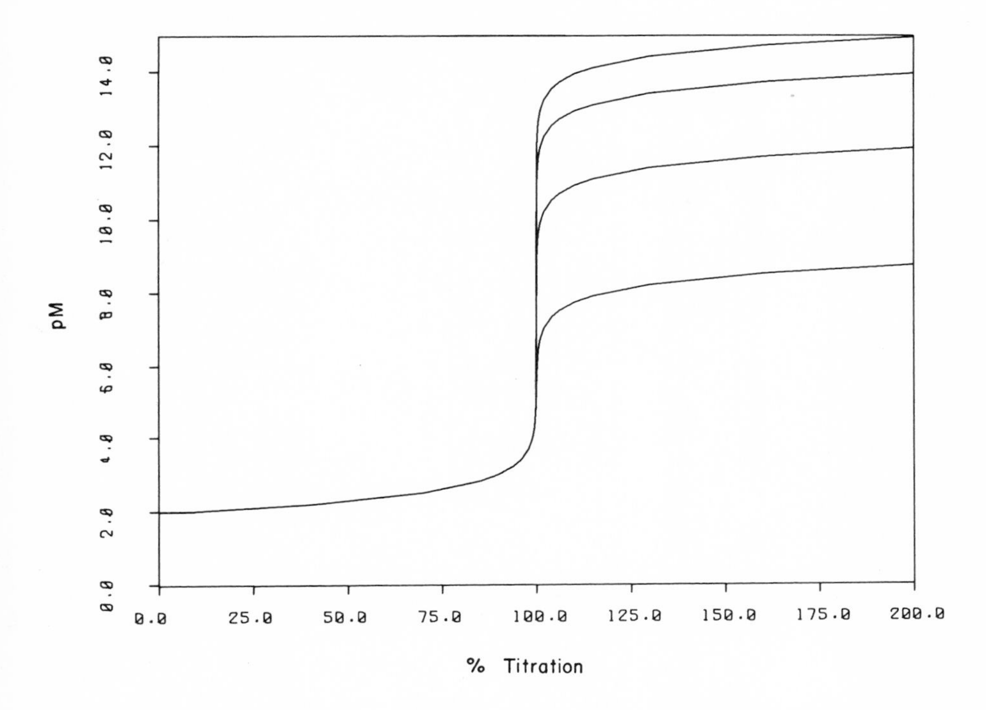

Fig. 9. EDTA titration curves as a function of pH: $[M^{z+}]_o = 0.01$ M; $K_f = 10^{15}$; pH = 5 (bottom), 7, 9, and 11 (top).

The ordinate is ΔpM = pM(101%) - pM(99%), that is, the change in pM between 1% beyond the equivalence point and 1% before the equivalence point. The initial metal ion concentration 0.1 M and the three curves were calculated using $K_f = 10^8$ (bottom), 10^{12}, and 10^{16} (top). To observe a ΔpM of at least 2, an EDTA titration of a 0.1 M metal with $K_a = 10^8$ would have to be buffered at about pH = 11. If $K_f = 10^{16}$, one could observe a ΔpM of 2 at pH 4.

The program NERNST stimulates a reversible metal, metal ion electrode immersed in an aqueous ammonia solution. If the metal ion complexes ammonia, the electrode potential will show a pH dependence (assuming that metal ion hydrolysis, hydroxide precipitation, and other complicating factors can be neglected). For the half-reaction, $M^{z+} + ze^- \rightleftarrows M$, the Nernst equation (at 25°C and assuming unit activity coefficients) is

$$E_{M^{z+},M} = E^{\circ}_{M^{z+},M} + \frac{0.059}{z} \log [M^{z+}].$$

But in aqueous ammonia most metal ions form ammine complexes:

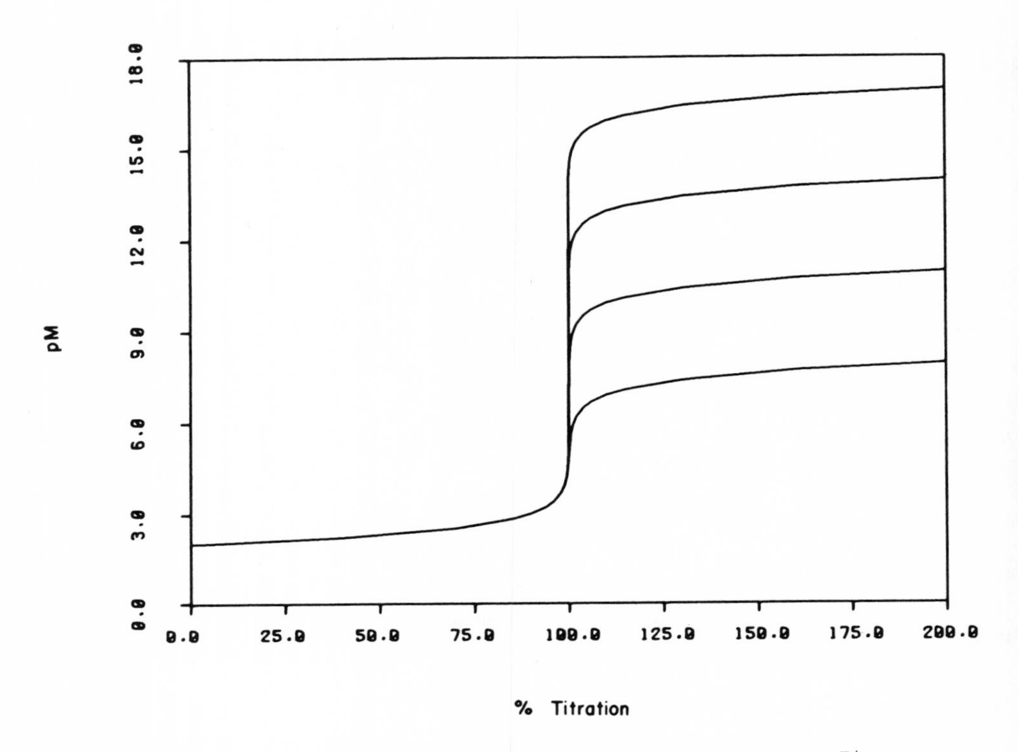

Fig. 10. EDTA titration curves as a function of K_f: pH = 11; $[M^{Z+}]_o$ = 0.01 M; $K_f = 10^8$ (bottom), 10^{11}, 10^{14}, and 10^{17} (top).

$$M^{Z+} + NH_3 \rightleftarrows M(NH_3)^{Z+} \qquad K_1$$

$$M(NH_3)^{Z+} + NH_3 \rightleftarrows M(NH_3)_2^{Z+} \qquad K_2$$

$$\vdots$$

$$M(NH_3)_{n-1}^{Z+} + NH_3 \rightleftarrows M(NH_3)_n^{Z+} \qquad K_n$$

where K_i is the stepwise formation constant for the complex, $M(NH_3)_i^{Z+}$. Therefore, the concentration of free metal ion is given by

$$[M^{Z+}] = \frac{C_m}{1 + K_1[NH_3] + K_1K_2[NH_3]^2 + K_1K_2K_3[NH_3]^3 + \ldots + K_1K_2K_3 \ldots K_n[NH_3]^n},$$

when C_m is the total metal ion concentration and $[NH_3]$ is the free ammonia concentration. The total ammonia concentration C_a is given by

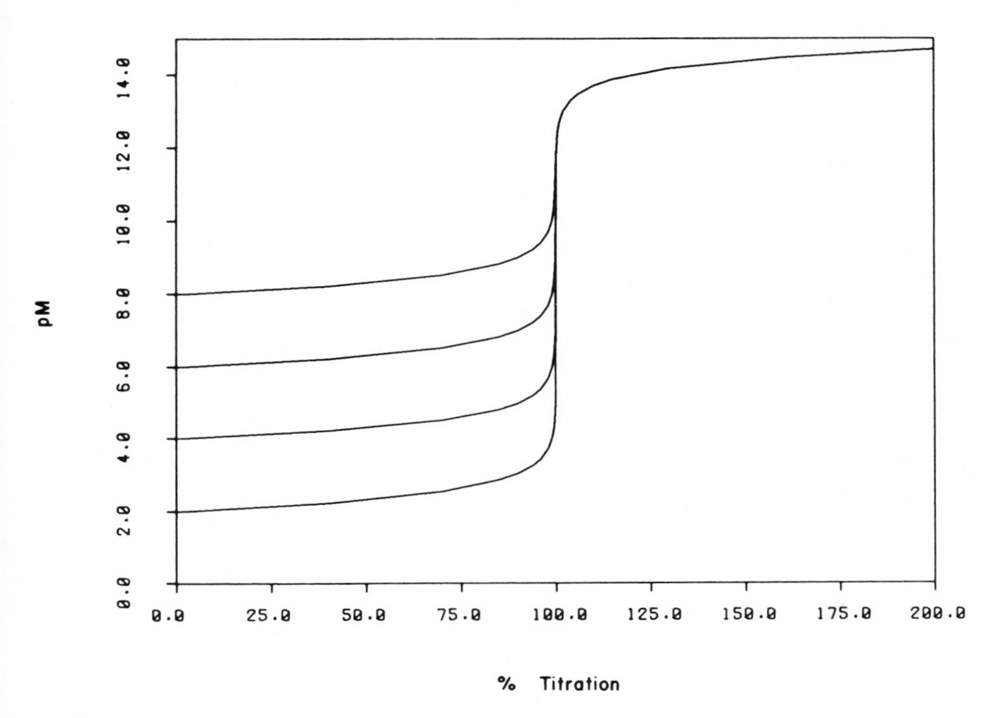

Fig. 11. EDTA titration curves as a function of $[M^{Z+}]_o$: pH = 10, $K_f = 10^{15}$, $[M^{Z+}]_o = 10^{-2}$ M (bottom), 10^{-4}, 10^{-6}, and 10^{-8} M (top).

$$C_a = [NH_3] + [NH_4^+] + \sum_{i=1}^{n} i[M(NH_3)_i].$$

If $C_m \ll C_a$, then

$$C_a \approx [NH_3] + [NH_4^+] = [NH_3]\left[\frac{K_a + [H^+]}{K_a}\right]$$

where K_a is the dissociation constant of NH_4^+. Thus, for a given pH, the free ammonia concentration is calculated first, then the free metal ion concentration, and finally, the electrode potential. Outputs from NERNST for Hg, Cd, and Zn electrodes are shown in Fig. 13. NERNST beautifully synthesizes three concepts in analytical chemistry: acid/base equilibria, complexation, equilibria, and the application of the NERNST equation. Students can be challenged to explain the shapes of the curves: why does the pH dependence cease after pH ≈ 10; why is there no pH dependence for Cd^{2+}

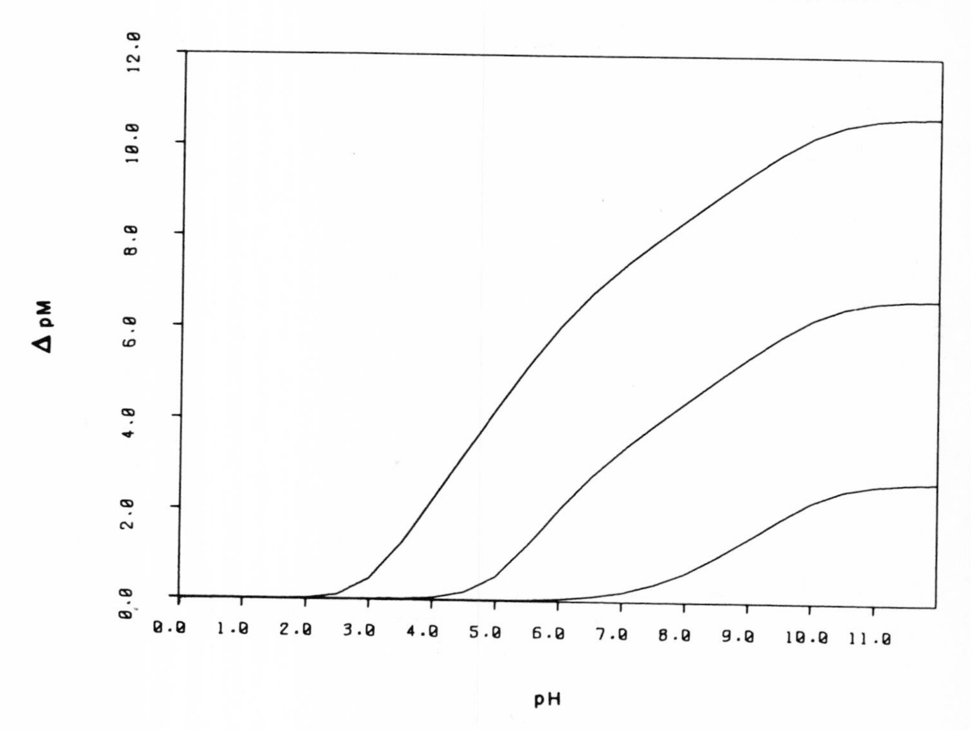

Fig. 12. EDTA titration feasibility plots; $\Delta pM = pM(101\%) - pM(99\%)$: $[M^{Z+}]_o = 0.1$ M; $K_f = 10^{16}$ (top), 10^{12}, and 10^{8} (bottom).

and Zn^{2+} below pH ≈ 6; and why does the $Hg^{2+}|Hg$ electrode show strong pH dependence until pH ≈ 10?

TITR is a data reduction program that has been used in the sophomore level analytical chemistry laboratory. The program computes the two thermodynamic dissociation constants of β-alanine:

$$H_2A^+ \rightleftarrows H^+ + HA \qquad K_1^o = \frac{a_{H^+} \times a_{HA}}{a_{H_2A^+}},$$

$$HA \rightleftarrows H^+ + A^- \qquad K_2^o = \frac{a_{H^+} \times a_{A^-}}{a_{HA}}.$$

The students collect data (pH vs volume of titrant) in the laboratory for the two titrations (β-alanine with strong acid and β-alanine with strong base). TITR prompts the students to enter the data, one pH-volume pair at a time,

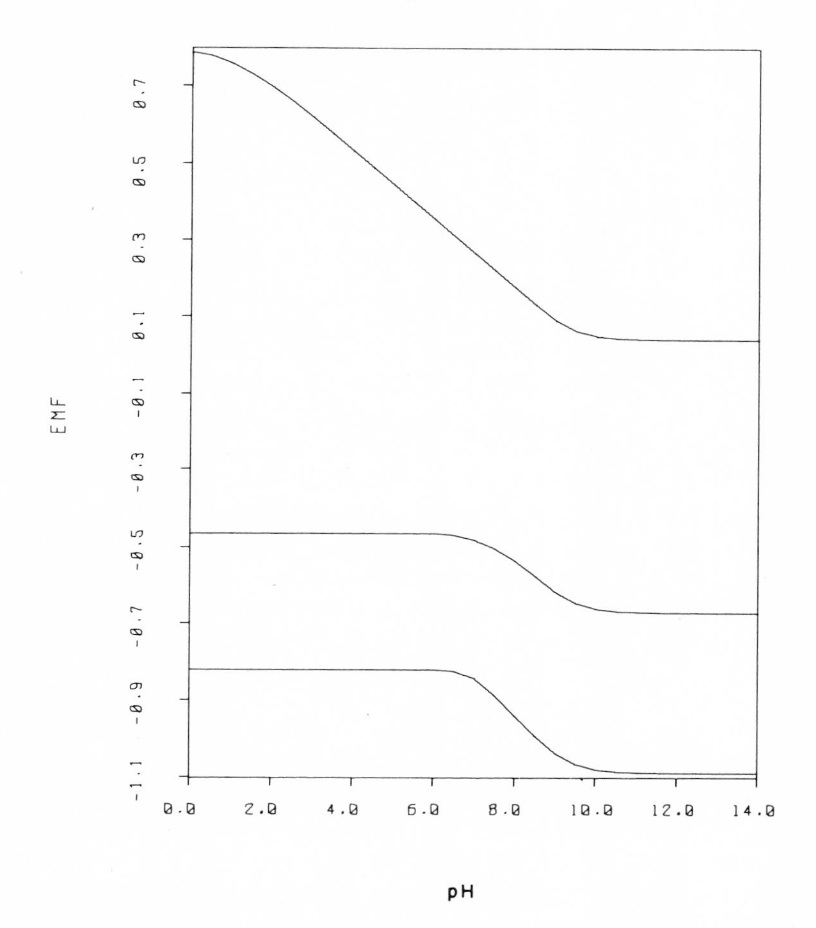

Fig. 13. Potential-pH curves for Hg^{2+}|Hg (top), Cd^{2+}| Cd, and Zn^{2+}|Cd, (bottom) electrodes in aqueous ammonia solution.

uses the limiting Debye-Hückel equation to approximate the activity coefficients, calculates the dissociation constant at each point, and then computes the average value and standard deviation of the dissociation constants.

Some of the simulation programs made available to students in undergraduate physical chemistry courses are described in Table 3.

The program GAS allows students to explore the van der Waals model of real gas behavior. The adjustable parameters include the two van der Waals constants a and b, and the temperature. The temperature must be greater than the critical temperature (8a/27Rb). The program also reads

TABLE 3

Physical Chemistry Programs

Name	Description
GAS	Calculates the compressibility factor of a van der Waals gas over the range $P_o \leq P \leq P_f$. Parameters: a, b, T
EQUIL	Calculates the equilibrium constant of a gas phase equilibrium system by finding the minimum of the total free energy as a function of extent of reaction. Parameters: temperature, stoichiometry and chemical potentials
BOX	Calculates eigenvalues of a particle in a finite and infinite square well potential. Parameters: mass of the particle, width of the well, magnitude of the well potential
ENTROPY	Calculates the molar entropy of a gas from spectroscopic data. Parameters: number of atoms in the molecule, linearity, mass and coordinates of each atom relative to an arbitrary origin, temperature, pressure, symmetry factor, multiplicity, and the fundamental vibration frequencies
NMR	Simulates AB, AB_2, ABX, A_2X_2, and A_2B_3 NMR spectra. Parameters: chemical shifts and coupling constants
ABC	Simulates an ABC NMR spectrum by diagonalizing the spin Hamiltonian. Parameters: three chemical shifts and three coupling constants.
HMO	Solves Hückel molecular orbital problems

the initial and final pressures that define the desired range of pressures, $P_o \leq P \leq P_f$. The corresponding molar volumes are calculated using the Newton-Rapheson method to find the root of the cubic equation

$$F(V) = PV - RT + \frac{a}{V} - Pb - \frac{ab}{V^2} = 0.$$

The program increments the molar volume, calculates P from the linear equation

$$P = \frac{RT}{V - b} - \frac{a}{V^2},$$

and then computes the compressibility factor, $Z = PV/RT$. A table containing V, P, and Z is generated. Typical plotted output is shown in Fig. 4.

EQUIL computes the equilibrium constant for the gas phase reaction

$$aA + bB \rightleftarrows cC + dD, \qquad K_p = \frac{p_C^c p_D^d}{p_A^a p_B^d}$$

by minimizing the total free energy of the system as a function of extent of reaction [10]. Assuming ideal gas behavior and one atmosphere total pressure, we find

$$G_{tot} = n_A\mu_A + n_B\mu_B + n_C\mu_V + n_D\mu_D,$$

where n_i is the number of moles of species i and

$$\mu_i = \mu_i^{\circ} + RT \ln p_i.$$

If initially there are a moles of A, b moles of B, and zero moles of C and D, then, when the extent of reaction is α,

$$n_A = a(1-\alpha) \qquad \mu_A = \mu_A^{\circ} + RT \ln\{a(1-\alpha)/w\},$$

$$n_B = b(1-\alpha) \qquad \mu_B = \mu_B^{\circ} + RT \ln\{b(1-\alpha)/w\},$$

$$n_C = c\alpha \qquad \mu_C = \mu_C^{\circ} + RT \ln\{c\alpha/w\},$$

$$n_D = d\alpha \qquad \mu_D = \mu_D^{\circ} + RT \ln\{d\alpha/w\},$$

where w = total moles = $(1-\alpha)(a+b) = \alpha(c+d)$.

EQUIL first evaluates G_{tot} for $0 \le \alpha \le 1$ in steps of 0.04. Then the iterative binary bisection method [11] is used to find the extent of reaction corresponding to the minimum in $G_{tot}(\alpha)$. Figure 14 illustrates the results for the dissociation of HI at 667°K. The program finds $\alpha_{MIN} = 0.2052$; $K_p = 1.67 \times 10^{-2}$. The same value can more easily be calculated using $\Delta G^{\circ} = -RT \ln K_p$. EQUIL provides an interesting additional approach to the equilibrium concept. It beautifully demonstrates that the equilibrium configuration of a chemical reaction corresponds to the minimum in total free energy.

BOX solves the one-dimensional particle in a finite potential well problem. The allowed energy levels E of the particle are the roots of the transcendental equation [12]

$$\tan\left(\frac{a(2mE)^{1/2}}{\hbar}\right) = \frac{2(V_o - E)^{1/2} E^{1/2}}{2E - V_o},$$

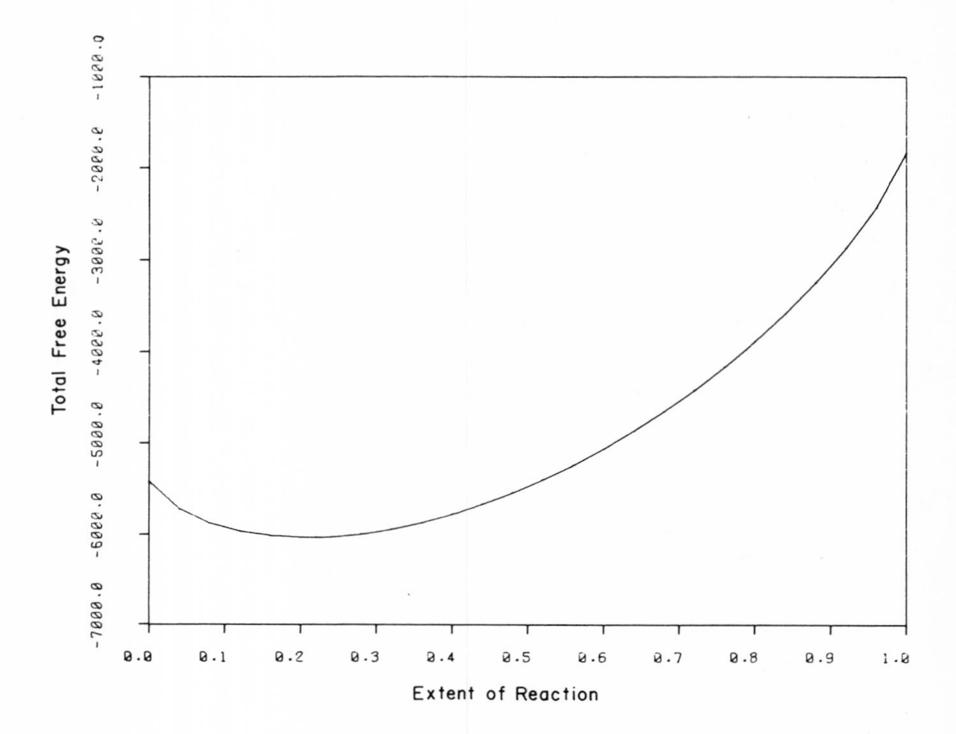

Fig. 14. Total free energy as a function of extent of the reaction $2HI \rightleftarrows H_2(g) + I_2(g)$, $T = 667^\circ K$.

where a is the width of the well, m is the mass of the particle, and V_o is the height of the potential well. The left- and right-hand sides of this function are plotted in Fig. 15 for a = 4 Å, $m = m_e = 9.1 \times 10^{-28}$ g, and V_o = 100 eV. The allowed eigenvalues (energy levels) of the electron are the intersections of the tangent lines with the right-hand side.

BOX finds the first six energy levels and compares them with the infinite potential well solution, $E_n = n^2h^2/8ma^2$, n = 1, 2, 3, ... For example, for a = 4 Å, $m = m_e$, V_o = 100 eV:

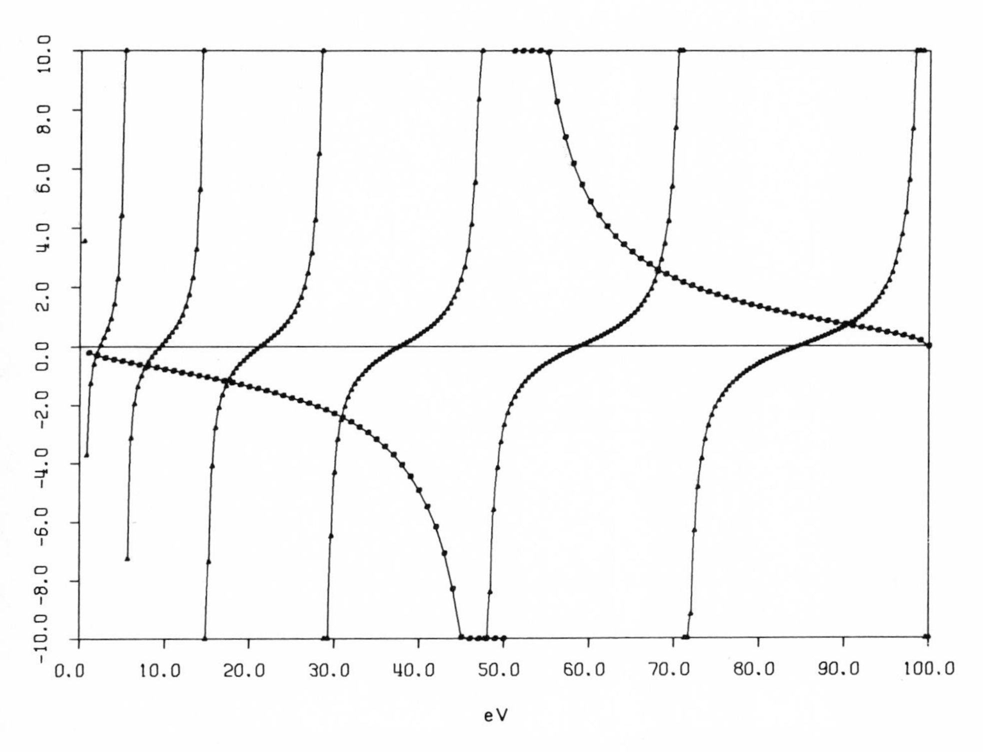

Fig. 15. Plot for graphical determination of the eigenvalues of a particle in a finite potential well (see text).

	FINITE BOX		INFINITE BOX
NUMBER	ENERGY IN EV	ENERGY IN ERG	ENERGY IN EV
1	1.94	3.12E-12	2.35
2	7.78	1.25E-11	9.40
3	17.46	2.80E-11	21.15
4	30.88	4.95E-11	37.60
5	47.88	7.67E-11	58.76
6	68.08	1.09E-10	84.61

By allowing m and/or a to increase, the student can observe the collapse of the energy level spacing; by allowing V_0 to increase, he can observe the approach to the infinite well solution. Figure 16 demonstrates the latter

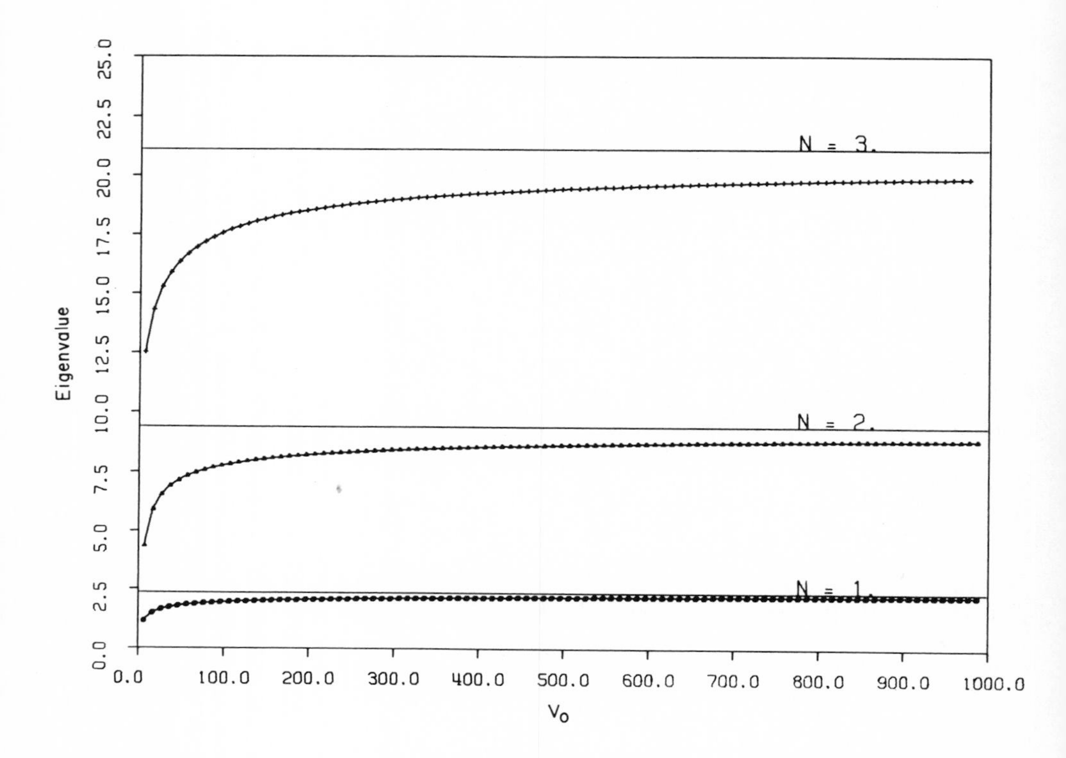

Fig. 16. The first three eigenvalues of an electron in both infinite (horizontal lines) and finite potential wells.

phenomenon for the first three eigenvalues.

ENTROPY calculates three of the components (translational, vibrational, and rotational) of the total entropy of a molecule in the gas phase from geometric and spectroscopic data [13]. For example, for the molecule $HCCl_3$, the program reads the following: number of atoms, 5; linear or nonlinear, nonlinear; temperature, 298 $^{\circ}K$; symmetry factor, 3; multiplicity of the electronic ground state, 1; the 3N-6 or 9 fundamental frequencies, 3019, 261, 261, 668, 368, 1216, 1216, 757, and 757; and the mass and coordinates of each atom,

MASS	X	Y	Z
1.008	0.0	0.0	1.093
12.001	0.0	0.0	0.0
35.453	1.467	0.847	-0.512
35.453	0.0	-1.694	-0.512
35.453	-1.467	0.847	-0.512

The program assumes the following: ideal gas behavior; translation, vibration, and rotation are mutually independent, so $Q_{tot} = Q_{trans} \cdot Q_{rot} \cdot Q_{vib}$ (Q = partition function); the molecule is in its ground electronic state; there is no internal rotation or other degrees of freedom; the classical expressions for translation and rotation; the harmonic oscillator approximation for vibration; and the effects of nuclear spins at low temperatures and anharmonicity at high temperatures can be neglected. The three moments of inertia are calculated using the rigid molecule method of Hirschfelder [14]. For $HCCl_3$ these are $I_{xx} = 2.62 \times 10^{-40}$, $I_{yy} = 2.62 \times 10^{-40}$, $I_{zz} = 5.07 \times 10^{-40}$ g cm^2. The program then prints the results: $S_{trans} = 40.2$, $S_{rot} = 25.2$, $S_{vib} = 5.31$, and $S_{tot} = 70.7$ entropy units. Using this program a student can discover the sensitivity of the three components of molecular entropy to geometry, temperature, vibrational modes, etc.

The program NMR uses closed-form expressions to simulate AB, AB_2 ABX, A_2X_2, and A_2B_3 NMR spectra [15]. The student enters a key to denote which of the five cases is of interest, and then the computer prompts him to enter the coupling constants and chemical shifts. The program prints the frequency and intensity of the allowed transitions and then, assuming overlapping Lorentzian bands of constant width at half-maximum intensity, prints a "typewriter art" spectrum at the terminal. A Calcomp version of the output is shown in Fig. 17a,b for the AB and X parts of an ABX spectrum ($\delta_A = 97$, $\delta_B = 103$, $\delta_X = 400$, $J_{AB} = 4.0$, $J_{AX} = 2.0$, $J_{BX} = 6.0$ Hz). By manipulating the chemical shifts and coupling constants, a student can quickly and painlessly discover how the bands in an NMR spectrum are affected by changes in these parameters.

The program ABC computes the 14-line ABC NMR spectrum as a function of the three chemical shifts (δ_A, δ_B, δ_C) and the three coupling constants (J_{AB}, J_{AC}, J_{BC}). The program uses a matrix diagonalization routine to solve the nuclear spin Hamiltonian [15, 16].

HMO is a program that solves simple Hückel π-molecular orbital problems [17]. The program reads the $n \times n$ secular determinant, where n is the number of atoms in the molecular plane. For example for 3-methylene-2,4-pentadiene

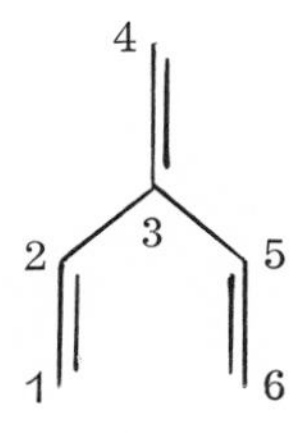

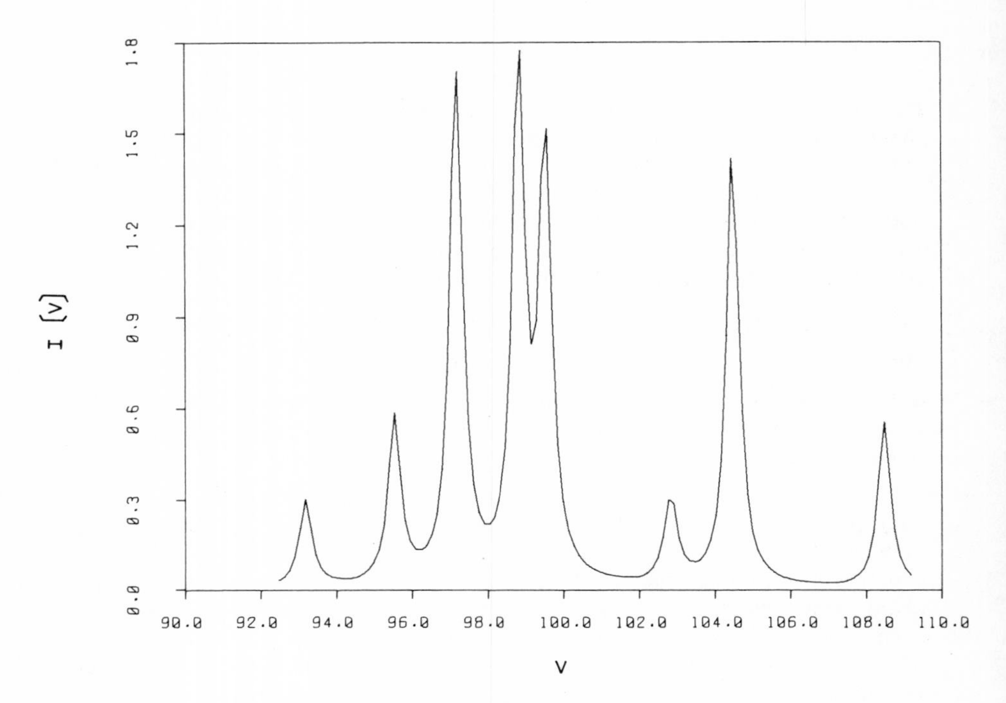

Fig. 17a. AB part of an ABX NMR spectrum. Chemical shifts: $\delta_A = 97$, $\delta_B = 103$, $\delta_X = 400$. Coupling constants: $J_{AB} = 4.0$, $J_{AX} = 2.0$, and $J_{BX} = 6.0$ Hz.

the secular determinant (zero overlap) is

$$\begin{vmatrix} 0 & 1 & 0 & 0 & 0 & 0 \\ 1 & 0 & 1 & 0 & 0 & 0 \\ 0 & 1 & 0 & 1 & 1 & 0 \\ 0 & 0 & 1 & 0 & 0 & 0 \\ 0 & 0 & 1 & 0 & 0 & 1 \\ 0 & 0 & 0 & 0 & 1 & 0 \end{vmatrix} = 0.$$

The roots of the characteristic equation are obtained by matrix diagonalization. The six roots are ± 1.93, ± 1.00, and ± 0.518. Thus, the energy level diagram for the molecule is

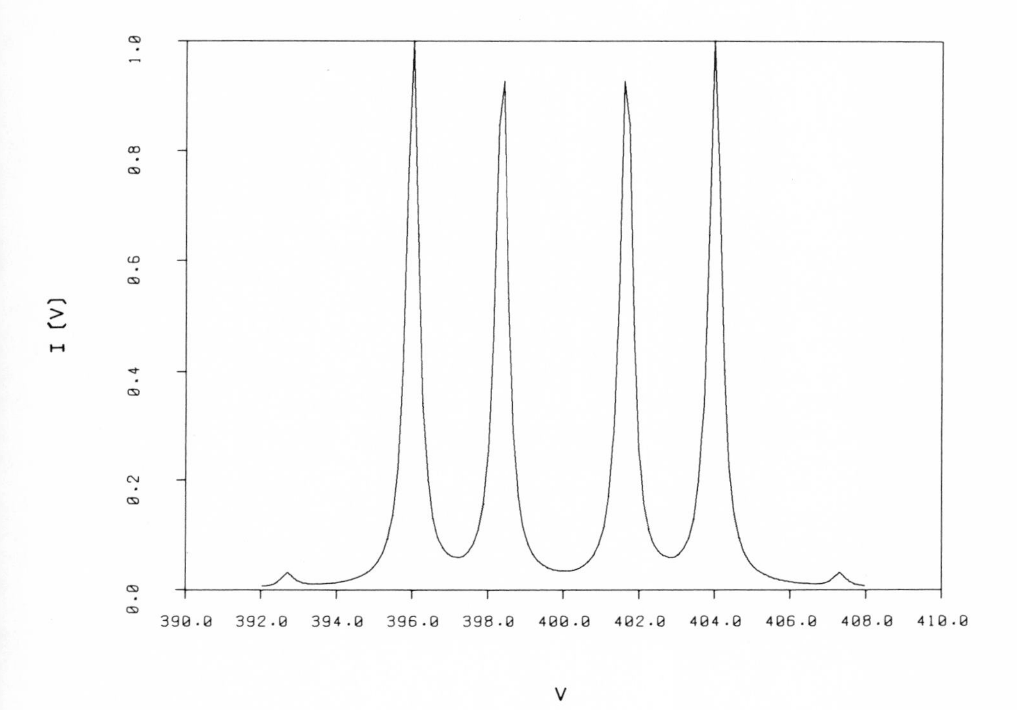

Fig. 17b. X part of an ABX NMR spectrum. Chemical shifts: $\delta_A = 97$, $\delta_B = 103$, $\delta_X = 400$. Coupling constants: $J_{AB} = 4.0$, $J_{AX} = 2.0$, and $J_{BX} = 6.0$ Hz.

Energy

E_6		$\alpha - 1.932\beta$
E_5		$\alpha - 2.000\beta$
E_4		$\alpha - 0.518\beta$
E_3	XX	$\alpha + 0.518\beta$
E_2	XX	$\alpha + 1.000\beta$
E_1	XX	$\alpha + 1.932\beta$

where α is the coulomb integral and β is the exchange integral. The matrix diagonalization routine also returns the n eigenvectors which may be used to calculate atomic electron densities, bond orders, and free valencies. The program calculates the following π-bond orders:

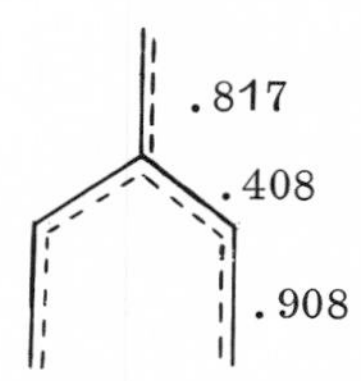

Students can use this program to discover the bonding and electron charge distribution in planar unsaturated molecules and how these are influenced by ionization, substituents, introduction of heteroatoms, etc.

Table 4 contains a description of some of the Fortran IV programs that are used to solve typical chemical numerical analysis problems. The matrix inversion routine, MATINV, uses the standard elimination method [18]. The largest eligible element in the matrix is used for the pivot element at each step in the elimination procedure. If the current pivot is less than an arbitrary tolerance the elimination procedure is terminated and a value of zero is assigned to the determinant. ITINV applies an iterative Newton-Rapheson procedure for improving the inverse. SIMQ calls MATINV to invert the coefficient matrix to solve linear systems of equations,

TABLE 4

Numerical Analysis Programs

Name	Description
MATINV	Matrix inversion routine
ITINV	Iterative matrix inversion routine
SIMQ	Solves systems of linear simultaneous equations by coefficient matrix inversion (MATINV)
POLREG	Linear and polynomial regression analysis
NONLIN	Nonlinear regression analysis (curve fitting)
NEWTON	Solves systems of nonlinear simultaneous equations using the Newton-Rapheson method
SECANT	Solves systems of nonlinear simultaneous equations using the secant method
RUNGE	Solves systems of simultaneous differential equations
JACOBI	Matrix diagonalization routine (Jacobi method)

$$AX = B$$
$$AA^{-1}X = A^{-1}B.$$
$$X = A^{-1}B$$

POLREG uses the standard least squares method [19] to perform linear and polynomial regression analysis. There are two weighting options, unit weighting and "statistical weighting," $w_i = 1/\sigma_i \approx 1/y_i^2$ [19b]. Here σ_i is the observed variance in y_i. The program prints a table of observed abscissas and ordinates, calculated ordinates and the difference between observed and calculated ordinates, the variance of the fit, the regression coefficients, and

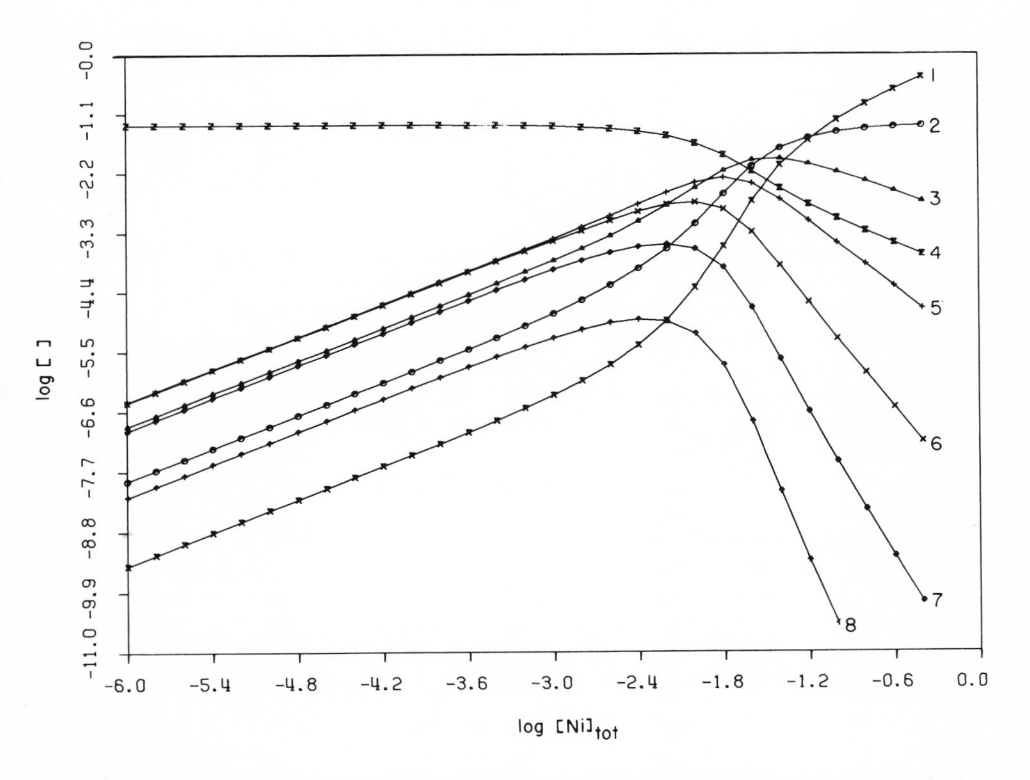

Fig. 18. Solution of the $Ni(NO_3)_2$-NH_4NO_3-KOH-H_2O system as a function of total Ni^{2+}; $C_{KOH} = 0.05$ M, $C_{NH_4NO_3} = 1.5$ M, T = 298°K. Legend: X (1) = $[Ni^{2+}]$; ⊖ (2) = $[Ni(NH_3)^{2+}]$; ▲ (3) = $[Ni(NH_3)_2^{2+}]$; Z (4) = $[NH_3]$; + (5) = $[Ni(NH_3)_3^{2+}]$; X (6) = $[Ni(NH_3)_4^{2+}]$; ◆(7) = $[Ni(NH_3)_5^{2+}]$; ↑ (8) = $[Ni(NH_3)_6^{2+}]$.

the standard deviations of the regression coefficients. NONLIN performs iterative Gaussian nonlinear regression analysis [19b]. The user writes a subroutine containing the partial derivatives that define the Jacobian coefficient matrix. NONLIN has been used to solve a number of curve-fitting problems. NEWTON and SECANT use the Newton-Rapheson and secant methods [20], respectively, for solving systems of simultaneous nonlinear equations. For example, the system $Ni(NO_3)_2$-NH_4NO_3-KOH-H_2O contains (at one level of approximation) eleven species: $Ni(NH_3)_i^{2+}$, i = 0, 1, 2, ... 6; H^+, OH^-, NH_3, and NH_4^+. Figure 18 demonstrates the concentration of free NH_3 and the seven Ni^{2+} species as a function of total Ni^{2+}. RUNGE uses a fourth-order Runge-Kutta algorithm [21] to solve systems of simultaneous differential equations. RUNGE has been used to simulate complex kinetic schemes [22] to approximate solutions to Schrödinger harmonic oscillator problems and to calculate the trajectories of the three hydrogen atoms in the exchange reaction [23]

$$H_A + H_B - H_C \rightarrow H_A - H_B + H_C.$$

JACOBI uses the Jacobi method [24] to diagonalize Hermitian (real and symmetric) matrices. This routine is also used by ABC and HMO (see Table 3).

The simulation, data reduction, and numerical analysis programs reside in a program library and can be accessed by students or faculty members. Students are encouraged to write their own programs whenever feasible. However, writing a program like ENTROPY is a major undertaking for most undergraduates. The objective is not to turn chemistry students into expert computer programmers, but rather to provide tools with which inquisitive students can discover physical and chemical phenomena.

D. The Numerical Methods Course

One of the most rewarding components of this work has been the development of the upper-division elective, "Numerical Methods in Chemistry" [25]. This course is taken by junior and senior chemistry majors, a few chemistry graduate students, and a few students from other departments. The course does not have an introductory computer science course as a prerequisite. However, many of the students have had some exposure to Fortran and/or PIL.

The objectives are twofold: that the students learn to write programs in Fortran IV, and that the students learn to write or use computer programs to solve some problems in chemistry that would not be attempted without a computer. The first third of the course is devoted to computer programming. The remaining time is spent surveying several topics in numerical analysis. The course is an applications course for chemists that condenses the information contained in several computer science, applied mathematics, and chemistry courses.

The lecture topics and computer programs discussed in the course are listed in Table 5.

PIL provides a delightful introduction to computer programming. Four lectures are adequate to introduce the fundamentals of numerical computer programming (input and output, branching, looping, the use of subroutines, etc.). The first two weekly assignments are to write a PIL program with at least one loop and one test, and a PIL program to perform linear regression analysis. The latter program includes the following output: intercept and slope; the standard deviation of the fit, intercept, and slope; the correlation coefficient; a table containing the observed independent and dependent variables, the calculated dependent variable, and the difference between observed and calculated dependent variables. Given a reliable time-sharing system, an adequate supply of remote terminals, and an interpretive language like PIL or BASIC, most students without previous programming experience can write moderately sophisticated data reduction and simulation programs in two or three weeks.

With this background two additional weeks are adequate for the introduction of Fortran IV programming. The next two weekly assignments are to write a Fortran program with READ, WRITE, DO, and IF statements, and a program with a SUBROUTINE or FUNCTION subprogram. The class discussion is primarily devoted to sample programs. Fortran is presented by example, and the text [26] contains 44 computer programs, 20 of which are documented in detail. The discussion does not get bogged down with detailed formatting options, etc. Efficient coding techniques are introduced whenever appropriate and students are warned about round-off and truncation errors.

The Pitt time-sharing system supports a Fortran IV compiler and text editor. However, many students prefer to use the WATFOR compiler [27] at one of the remote-job-entry stations. Given reasonably fast turn-around at the RJE station (a few minutes), novice Fortran programmers can write and debug programs more efficiently using RJE than time-sharing.

The remaining two-thirds of the course is an applications approach to several topics in numerical analysis. The emphasis is on presentation and application rather than derivation and discussion. The students are referred to standard numerical analysis texts for a more detailed treatment of any topic that interests them.

The first topic in the numerical methods part of the course is roots of equations. The quadratic and cubic closed-form solutions are reviewed. The cubic solution can be used to solve for the three components of the moment of inertia of a nonlinear molecule from an arbitrary set of Cartesian coordinates. Using the Hirschfelder method [14],

TABLE 5

Numerical Methods Course — Topics and Programs

Topics	Programs
Part I Programming	
PIL Programming	STATS — Average and standard deviation
	GAS — Compressibility of a van der Waals gas
	QUAD — Quadratic equation solver
	LOCATE — Locate a root of an equation
	NEWT — Newton-Rapheson method for finding roots of equations
Fortran IV Programming	RADIO — Overlapping radiodecay curves
	BINARY — Binary bisection method for finding roots of equations
	SUBS — Programs illustrating the use of FUNCTION and SUBROUTINE subprograms
Part II Numerical Methods	
Roots of Equations	
Closed form (quadratic and cubic)	SUBROUTINE DCUBIC
Binary bisection method	BINBIS, BOX
Newton-Rapheson method	NEWRAP
Secant method	SECNT
Systems of Linear Equations	
Cramer's rule	
Gauss-Seidel method	SEIDEL
Gaussian triangularization	
Gauss-Jordan elimination	SUBROUTINE GAUJOR
Matrix inversion	SUBROUTINE MATINV
Regression Analysis	
Linear	LINEAR
Polynomial	POLREG

TABLE 5 (continued)

Topics	Programs
The Forsythe method	
Curve smoothing	
Nonlinear least squares analysis	NONLIN
Nonlinear Systems of Equations	
Successive approximations	APPROX
Newton-Rapheson method	NEWTON
Secant method	SECANT
Numerical Integration	
Trapezoidal rule	TRAP
Simpson's rule	SIMP
Cote's numbers	
Gaussian quadrature	LEGDRE
Monte-Carlo techniques	MONTE
Differential Equations	
Euler's method	EULER
Runge-Kutta algorithms	RUNKUT
Predictor corrector algorithms	PRECOR
Systems of differential equations	SYSDQ
Higher-order differential equations	HDIFFQ
Eigenvalues and Eigenvectors	
Evaluation by determinants	
The Jacobi method	JACOBI
Simulation of NMR spectra	ABC
Hückel molecular orbital theory	HMO
Miscellaneous Topics	
Graphics	CALCMP, PIGS, PLOT
Computer-assisted instruction	
Mini computers	
Information retrieval	IR
Artificial Intelligence	
Programming Considerations	

$$\begin{vmatrix} A-I_{xx} & -D & -E \\ -D & B-I_{yy} & -F \\ -E & -F & C-I_{zz} \end{vmatrix} = 0,$$

where

$$A = \Sigma m_i(Y_i^2 + Z_1^2) - \frac{1}{M}(\Sigma m_i Y_i)^2 - \frac{1}{M}(\Sigma m_i Z_m{}^2),$$

$$B = \Sigma m_i(X_i^2 + Z_i^2) - \frac{1}{M}(\Sigma m_i X_i)^2 - \frac{1}{M}(\Sigma m_i Z_i^2),$$

$$C = \Sigma m_i(X_i^2 + Y_i^2) - \frac{1}{M}(\Sigma m_i X_i^2 - \frac{1}{M}(\Sigma m_i Y_i^2),$$

$$D = \Sigma m_i X_i Y_i - \frac{1}{M}(\Sigma m_i X_i)(\Sigma m_i Y_i),$$

$$E = \Sigma m_i X_i Z_i - \frac{1}{M}(\Sigma m_i X_i)(\Sigma m_i Z_i),$$

$$F = \Sigma m_i Y_i Z_i - \frac{1}{M}(\Sigma m_i Y_i)(\Sigma m_i Z_i),$$

$$M = \Sigma m_i.$$

The binary bisection method [11] is useful for finding the roots of a transcendental function, for example, the eigenvalues of a particle in a finite box (see Figs. 15, 16). The Newton-Rapheson and secant methods [20] are covered next, and sample Fortran programs are discussed.

The next topic is methods of solving linear systems of simultaneous equations. Cramer's rule [28] is reviewed. The iterative Gauss-Seidel method [29] is then introduced, followed by a discussion of the elimination methods: Gaussian triangularization, Gauss-Jordan total elimination, and matrix inversion [30]. The relevant Fortran programs and subroutines are discussed in detail. One application of these programs is to solve for the unknown concentrations of components in a mixture determined by spectrophotometry or mass spectroscopy.

Regression analysis, an important application of linear systems of equations, is considered next. Statistical weighting [19] is introduced and the standard least-squares linear and polynomial techniques are discussed. The Forsythe method [31], which uses orthogonal polynomials, is then covered. Then the convolution technique of curve smoothing [32] is illustrated. This topic concludes with a detailed discussion of Gaussian nonlinear least-squares analysis [19]. This involves the use of an iterative least-squares technique to obtain parameters of a model that represents a chemical system.

For example, the four rate constants of the kinetic system

$$A \underset{k_4}{\overset{k_1}{\rightleftarrows}} B \underset{k_3}{\overset{k_2}{\rightleftarrows}} C, \qquad A(0) = 1,\ B(0) = C(0) = 0,$$

can be obtained from observed data, B(t), from [33]

$$B(t) = \frac{k_1 k_3}{\lambda_2 \lambda_3} + \left\{\frac{k_1(k_3 - \lambda_2)}{\lambda_2(\lambda_2 - \lambda_3)}\right\} e^{-\lambda_2 t} + \left\{\frac{k_1(\lambda_3 - k_3)}{\lambda_3(\lambda_2 - \lambda_3)}\right\} e^{-\lambda_3 t},$$

where

$$\lambda_2 = \tfrac{1}{2}(p + q), \qquad \lambda_3 = \tfrac{1}{2}(p - q),$$

$$p = k_1 + k_2 + k_3 + k_4, \qquad q = [p^2 - 4(k_1k_2 + k_3k_4 + k_1k_3]^{1/2}.$$

With adequate data [B(t) of sufficient precision], the iterative technique may converge on a set of rate constants that provide the best fit to the data in the least-squares sense.

The next topic is methods of solving systems of nonlinear simultaneous equations. For example, a kinetic study of the Ni^{2+}-ethylenediamine-H_2O system required equilibrium concentrations of all species as a function of analytical concentration and temperature [34]. The Newton-Rapheson and secant [20] methods are discussed and documented Fortran programs are used to solve similar nonlinear systems. Figure 18 shows part of the solution to the $Ni(NO_3)_2$-NH_4NO_3-KOH-H_2O system. The problem is to determine the equilibrium concentration of all species at 298°K, total KOH = 0.05M, total NH_4NO_3 = 1.5 M, as a function of total $Ni(NO_3)_2$. The equilibrium concentrations of NH_3 and $Ni(NH_3)_i^{2+}$, i = 0, 1, 2, ... 6 are plotted.

Numerical integration is the next topic. The following numerical methods are described: the trapezoidal rule, Simpson's rule, the use of Cote's numbers, Gaussian quadrature, and the Monte-Carlo technique [35]. Students write or adapt programs to solve error functions, gamma functions, fugacities of real gases, coefficients of Fourier expansions, topics from experimental heat capacity data, theoretical heat capacities from the Debye-Einstein equation, black body integrals, etc.

This leads to the numerical solution of differential equations and systems of differential equations. Three methods are presented: Euler's method, the Runge-Kutta algorithms, and the predictor-corrector technique [36]. Students use these methods to solve differential equations, simulate kinetic schemes, make trajectory calculations [23] and extend them to solve higher-order systems, for example, harmonic oscillator problems and solutions to the Schrödinger equation for a vibrating diatomic molecule.

The last numerical methods topic is the eigenvalue-eigenvector problem. The solution by determinants is described for low-order matrices. The

Jacobi method of matrix diagonalization by a series of orthogonal transformations [24] is discussed in some detail. The Givens-Householder [37] method is also introduced. Two specific applications are worked out in detail: the ABC NMR problem [15, 16], and simple HMO analysis [17].

If there is time, several additional topics are discussed. Most students are interested in using an incremental plotter, so the relevant information is presented. Some topics in non-numerical chemical applications are introduced: information retrieval, computer-assisted instruction, and artificial intelligence. Some time is given to a discussion of mini-computers and computer-instrument interfacing. And whenever appropriate, programming techniques and considerations are discussed.

In addition to weekly assignments, the students are responsible for a midterm and final project — to design fully documented programs. Some examples include: simulation of titration systems, radial distribution functions, thermodynamic relationships, crystal structure calculations, aqueous equilibrium systems, nonlinear least-squares programs, NMR and ESR spectral simulations, use of LACOON-III [38] to simulate complex NMR spectra, use of a CINDO (complete or incomplete neglect of differential overlap) program to obtain molecular orbitals [39], electron density contour plotting [40], simulation of kinetic schemes, numerical solution to the Schrödinger equation, Fourier analysis, and determination of bonding parameters from microwave data [41]. If a student is unable to decide on a project, he can be given many suggestions: a simulation or data reduction problem, a CAI lesson, or a few questions for the computer-generated repeatable examination system (see below). Most students document these programs beautifully.

There is an in-class midterm exam (programming), a take-home exam (numerical methods), and an in-class final exam (programming). No "cake" elective, this course attracts only well-motivated juniors and seniors. They realize that the course provides a unique opportunity to learn computer programming and several topics in numerical analysis from a chemical applications point of view.

E. Graphics

It has been demonstrated that simulation programs allow students to interact with chemical systems without being burdened with detailed numerical calculations. The effectiveness of these simulation routines is significantly enhanced if the results are displayed graphically. For example, consider the simulation of an ABX NMR spectrum (see Fig. 17). A table of transition frequencies and intensities can be calculated from closed-form expressions [15]. For example, for $\delta_A = 97$, $\delta_B = 103$, $\delta_C = 400$, $J_{AB} = 4$, $J_{AX} = 2$, and $J_{BX} = 6$ Hz (δ = chemical shift, J - coupling constant):

FREQUENCY	INTENSITY	ORIGIN
-200.00	0.0	*
93.17	0.293	A
95.53	0.553	A
97.17	1.707	A
98.83	1.707	B
99.53	1.447	A
102.83	0.293	B
104.47	1.447	B
108.47	0.553	B
392.70	0.026	X*
396.00	1.000	X
398.36	0.974	X
401.64	0.974	X
404.00	1.000	X
407.30	0.026	X*

The transitions designated * and X* are combination (forbidden) transitions. Figure 17 was drawn by assuming a Lorentzian band of constant half-width at half-maximum intensity (0.2 Hz) for each transition and then summing

$$I(\nu) = \frac{I_{0j}\delta^2}{\delta^2 + (\nu - \nu_{0j})^2},$$

where I_{0j} and ν_{0j} are the intensity and frequency of the jth transition and $\delta = 0.2$ Hz. Clearly, a student can more quickly develop a "feel" for the ABX system by watching bands separate, increase in intensity, etc., by observing graphical rather than tabulated numeric output.

Four approaches to graphics are discussed: line printer or teletype graphics ("typewriter art"); the use of an incremental plotter, e.g., the Calcomp plotter; the Culler-Fried interactive graphics system [3]; and PIGS, Pitt's Interactive Graphics System.

The least expensive graphics facility is the teletype or line printer. Figures 19 and 20 are typical "typewriter art" plots. Figure 19 is a simulated two-component radiodecay system:

$$A_{tot}(t) = A_1(t) + A_2(t),$$

$$A_{tot}(t) = A_{10}\exp(-0.693t/\tau_1) + A_{20}\exp(-0.693t/\tau_2),$$

where $A_{tot}(t)$ is the observed total activity at time t, A_{10} and A_{20} are the respective initial activities ($t = 0$), and τ_1 and τ_2 are the corresponding half-lives. Here $A_{10} = 2000$, $A_{20} = 1000$, $\tau_1 = 60$, and $\tau_2 = 120$. Figure 20

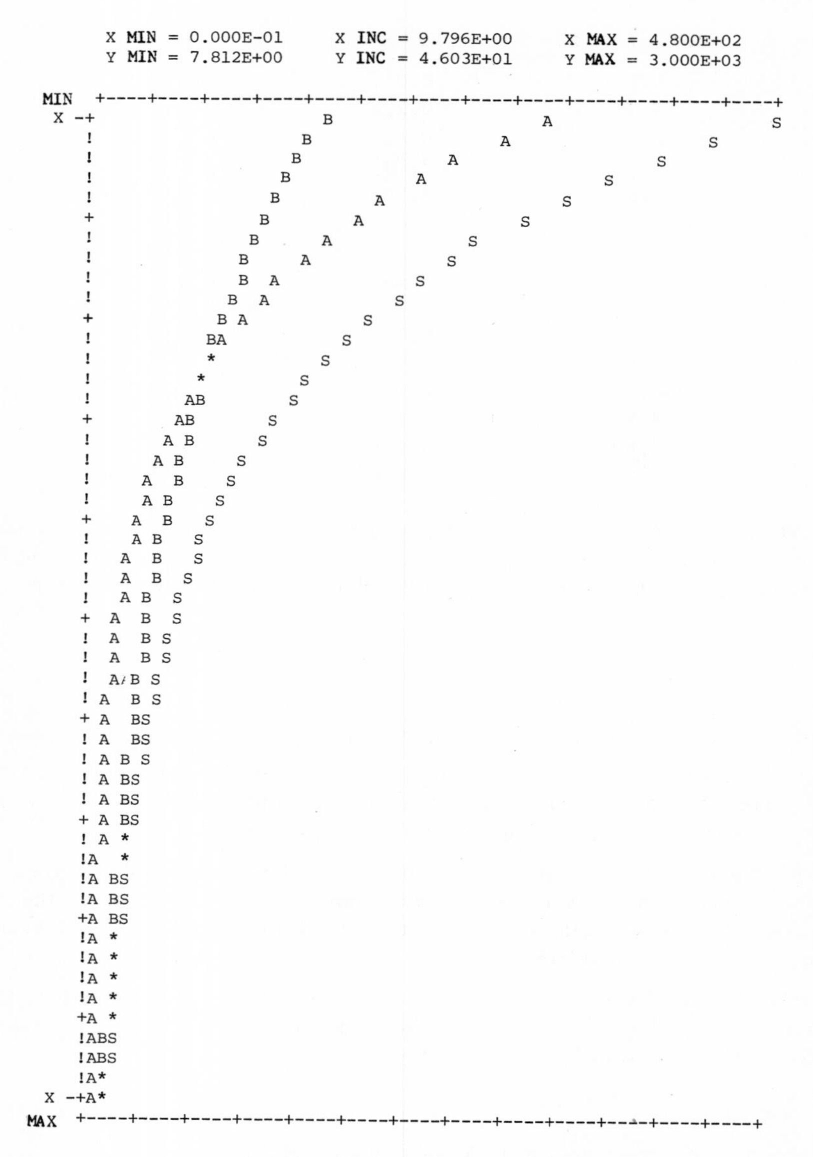

Fig. 19. Two component radiodecay, $S(t) = A(t) + B(t)$, where t is time, $A(t)$ is the activity of isotope A, and $B(t)$ is the activity of isotope B.

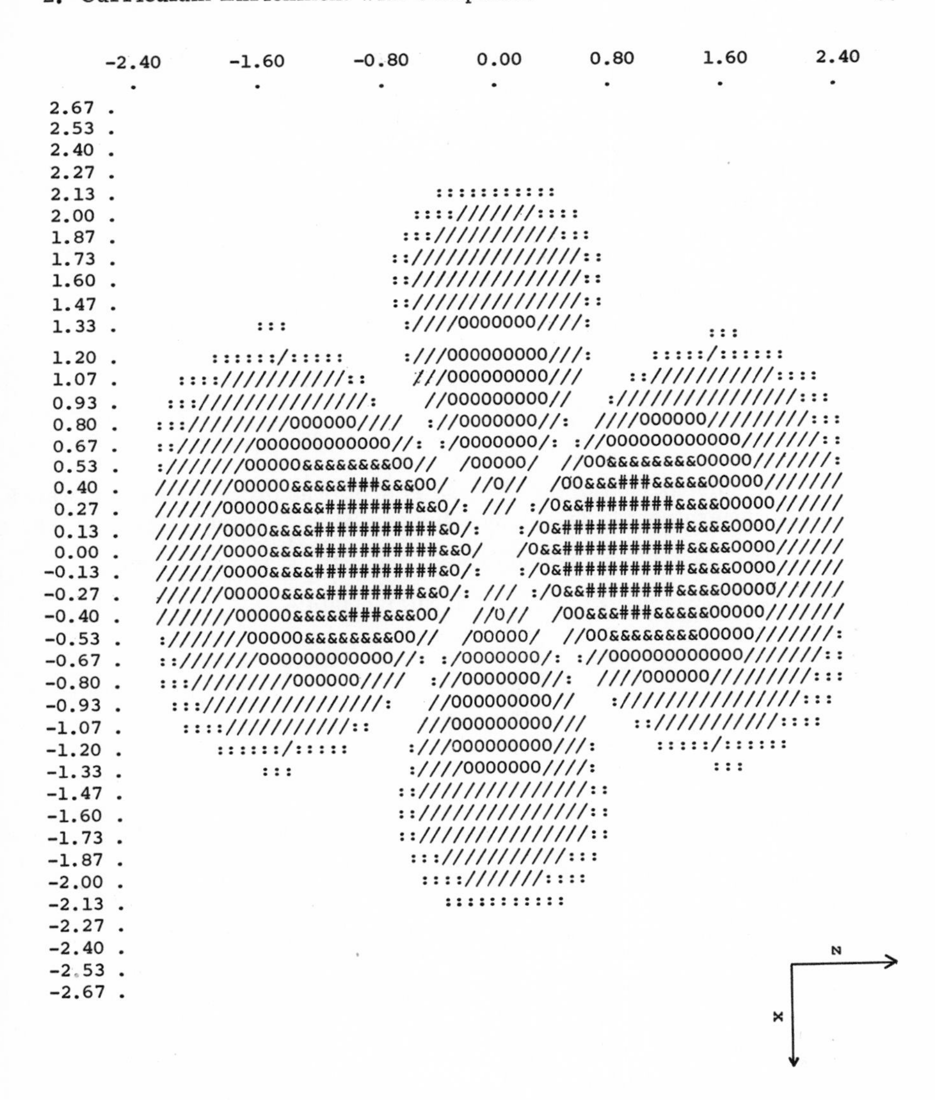

Fig. 20. The $3d_{z^2}$ orbital in the xz plane (see text for legend).

is an electron density contour plot for the $3d_{z^2}$ orbital. The program uses the hydrogen-like orbital [40a]:

$$\psi_{3d_{z^2}} = \frac{1}{81\sqrt{6\pi}}\left(\frac{Z}{a_o}\right)^{3/2} \sigma^2 e^{-\sigma/3} \quad (3\cos^2\theta - 1),$$

where Z is the effective nuclear charge, $a_o = 0.529$ Å and $\sigma = Zr/a_o$. In the xz plane, $x = r \sin\theta$, $z = r\cos\theta$, and $r^2 = x^2 + y^2$. The program systematically varies x and y, computes r, $\cos\theta$, and then $\psi^2_{3d_{z^2}}$. The program uses the following symbols to represent electron density ranges:

Symbol	Blank	:	/	0	&	#
Range	0-.01	.01-.02	.02-.10	.10-.25	.25-.50	.50-1.0

The effective nuclear charge in this case is 4.0. The program that generated Fig. 20 also generates 1s, 2s, 2p, 3p, 4p, and $3d_{x^2-y^2}$ atomic orbitals; sp, sp^2, sp^3, dsp^2, and d^2sp^3 hybrid orbitals; and $1s_a \pm 1s_b$, $2s_a \pm 2s_b$, $2p_x \pm 2p_x$, $2p_y \pm 2p_y$, and $2s_a \pm 2p_x$ molecular orbitals.

The use of the line printer or teletype for graphics is adequate for many purposes. However, an incremental plotter provides significantly enhanced resolution. All of the figures in this chapter except Figs. 1-3, 19, 20, and 23 were drawn by Pitt's Calcomp plotter. The figures are curvilinear, i.e., consist of a number of straight line segments. For example, Fig. 21 is an electron density contour plot of a hydrogen-like $3d_{x^2-y^2}$ orbital. This is a two-parameter system involving the effective nuclear charge Z and the contour line of interest. The square of the $3d_{x^2-y^2}$ wave function is [40a]

$$\Psi^2(r, \theta, \varphi) = \frac{1}{2\pi 3^8}\left(\frac{Z}{a_o}\right)^3 \sigma^4 \exp\left(\frac{-2\sigma}{3}\right) \cos^2 2\varphi \sin^4\theta.$$

In the x-y plane $\sin\theta = 1.0$. It is readily seen that the modified wave function

$$\Psi_M^2(r, \theta, \varphi) = C(Z)r^4 \exp(-b(Z)r) \cos^2 s\varphi$$

has maximum value 1.0 when $C(Z) = (Ze/6a_o)^4$ and $b(Z) = 2Z/3a_o$, where $e = 2.718\ldots$. For $0 < r < r_{max}$ and $0 < \Psi_M^2 < 1$,

$$\varphi = \left[\tan^{-1}\left\{\frac{C(Z)r^4 \exp[-b(Z)r] - \Psi_M^2}{\Psi_M^2}\right\}^{1/2}\right]/2.$$

The procedure is to input values for Z and Ψ_M^2, and in one loop which varies r, calculate φ, x, and y ($x = r\cos\varphi$ and $y = r\sin\varphi$) for one-eighth of the

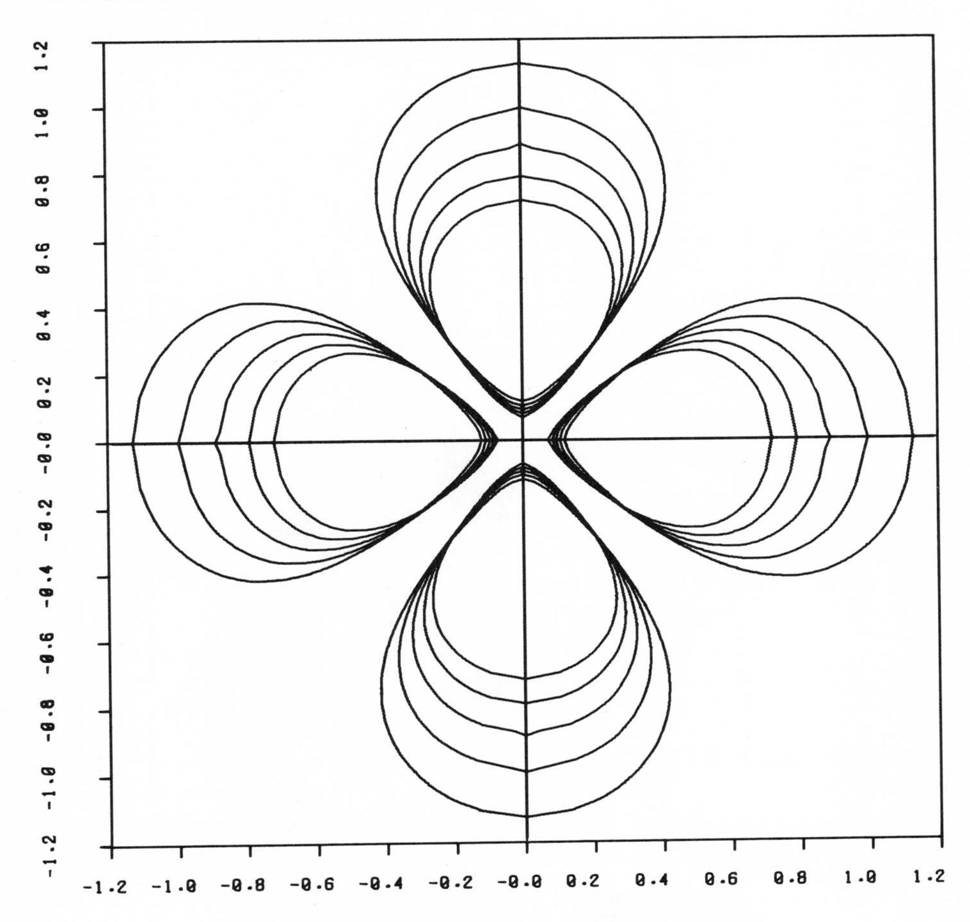

Fig. 21. The $3d_{z^2}$ orbital in the xy plane (see text).

$3d_{x^2-y^2}$ contour. The other seven segments are obtained from symmetry. In Fig. 21, $\Psi_M^2 = 0.1$ and Z is 7 (outermost contour), 8, 9, 10, and 11 (innermost contour). This procedure can be extended to plot other atomic orbitals, hybrid orbitals, and molecular orbitals [40].

Figure 22 is a three-dimensional plot of the total electron density of H_2CO [40b]. The electron density as a function of position was calculated from the occupied molecular orbitals using a Slater orbital basis set. The coefficients that define the LCAO-MOs were obtained using a CINDO program [39]. The plotting program has incorporated a hidden-line algorithm [42] to suppress the plotting of lines behind the peaks.

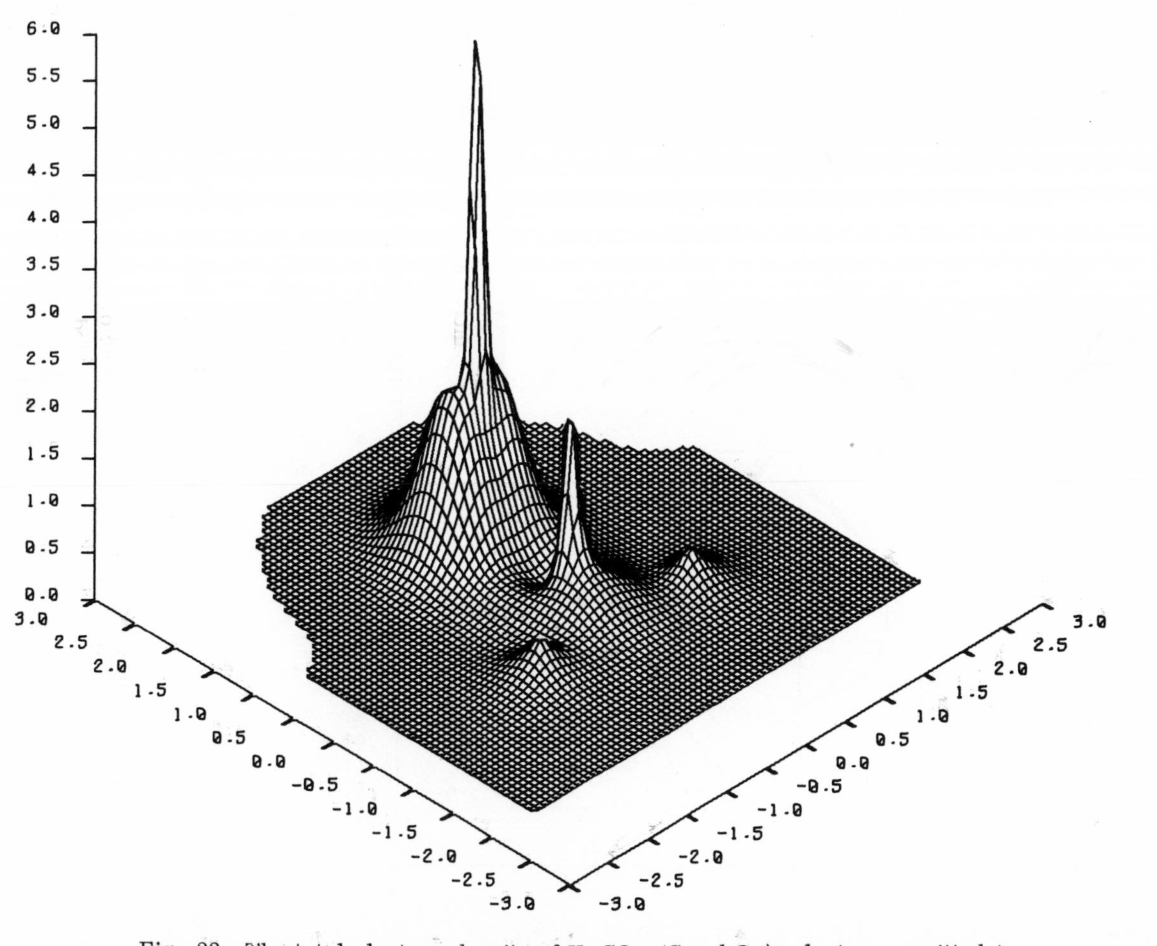

Fig. 22. The total electron density of H_2CO. (C and O 1s electrons omitted.)

The use of an incremental plotter clearly provides a substantial increase in resolution and appearance of graphic output. Another improvement in ease of use and convenience is attained with an interactive graphics system. Pitt became involved with interactive graphics via an NSF-sponsored network of ten universities tied to the University of California, Santa Barbara computing center. The UCSB system has evolved from the pioneering efforts of Culler and Fried [3]. A block diagram of the graphics terminal is shown in Fig. 23.

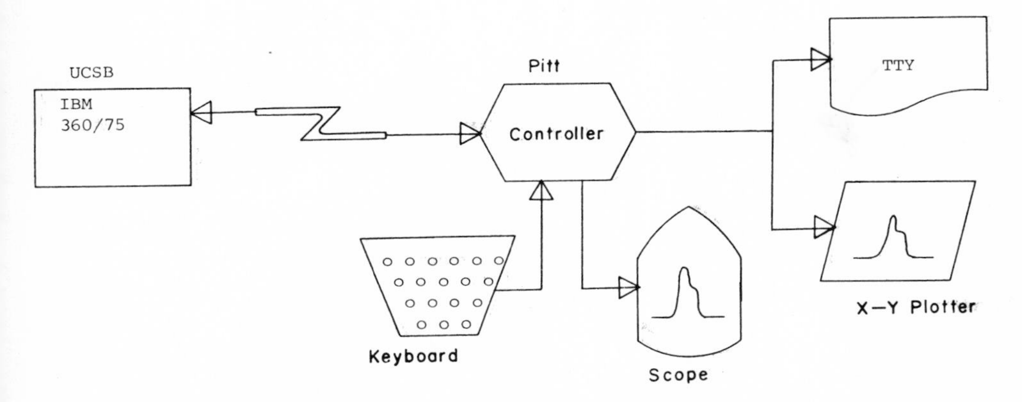

Fig. 23. Components of the interactive graphics terminal.

The user interacts with the system by pressing keys on the keyboard. Each time a key is pressed, information is transmitted to Santa Barbara, the operation is performed, and the results are available for display on the oscilloscope if desired. The teletype is used for program listings and numeric output. The plotter provides hard copy of information displayed on the scope. The programming language is a mathematically oriented, pseudo-assembly language. The entire instruction set is implemented on the keyboard, which has two sections — an operator section (LOAD, STORE, LOG, SIN, DISPLAY, etc.) and an operand section (standard character set). The NSF grant provided Pitt with the graphics terminal and access to the UCSB system for one year. During that time most of the simulation programs described above were implemented. Several other systems were also coded. For example Fig. 24 shows concentration-time profiles for the kinetic system

$$A \underset{k_4}{\overset{k_1}{\rightleftarrows}} B \underset{k_3}{\overset{k_2}{\rightleftarrows}} C.$$

This system is solved in closed form [33] for initial conditions $A(0) = a$, $B(0) = C(0) = 0$. The input to the program includes: k_1, k_2, k_3, k_4, and

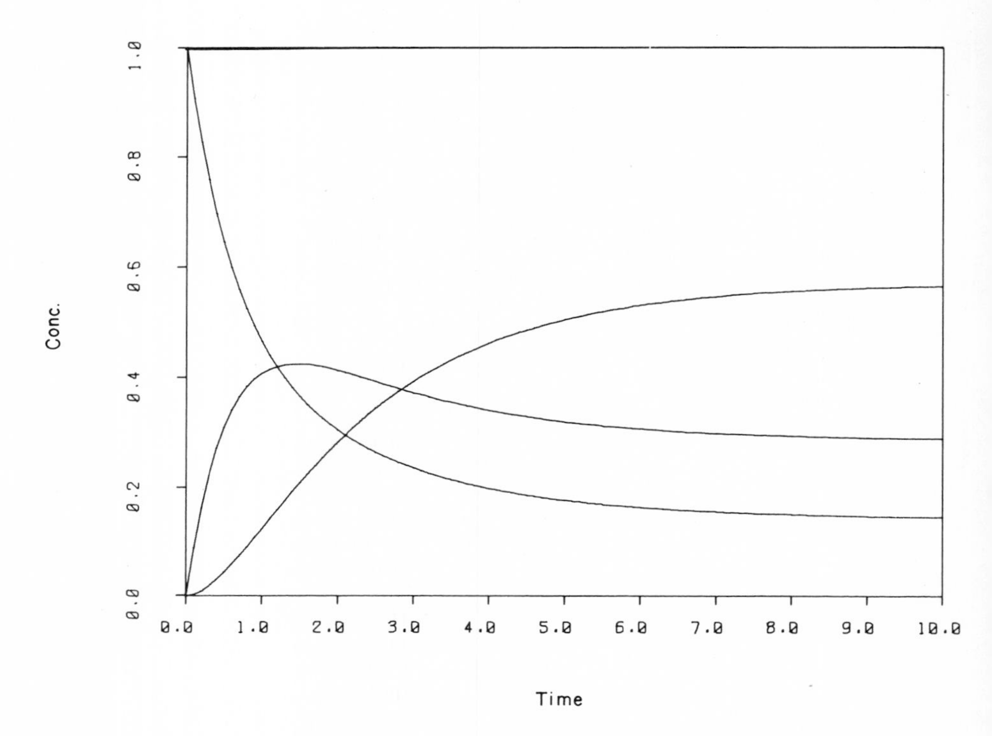

Fig. 24. Concentration-time profiles for the A ⇄ B ⇄ C system.

t_{eq}, where t_{eq} defines the abscissa ($0 \leq t \leq t_{eq}$). Two obvious applications are the "discovery" of the equilibrium properties, $K_1 = B_{eq}/A_{eq} = k_1/k_4$ and $K_2 = C_{eq}/B_{eq} = k_2/k_3$, and the conditions for which the steady state assumption ($dB/dt \approx 0$) is valid.

Figure 25 simulates the two-site NMR exchange system [43]. Given several simplifying conditions, the intensity of the NMR signal as a function of frequency is given by

$$I(\nu) = \frac{K(\nu_A - \nu_B)^2}{[\frac{1}{2}(\nu_A + \nu_B) - \nu]^2 + 4\pi^2\tau^2(\nu_A - \nu)^2(\nu_B - \nu)^2},$$

where K is a normalizing constant (taken as unity), τ is the lifetime, ν_A and ν_B are resonance frequencies of sites A and B, and ν is the frequency. If $2\pi\tau(\nu_A - \nu_B) > \sqrt{2}$, two lines result. The two peaks coalesce when

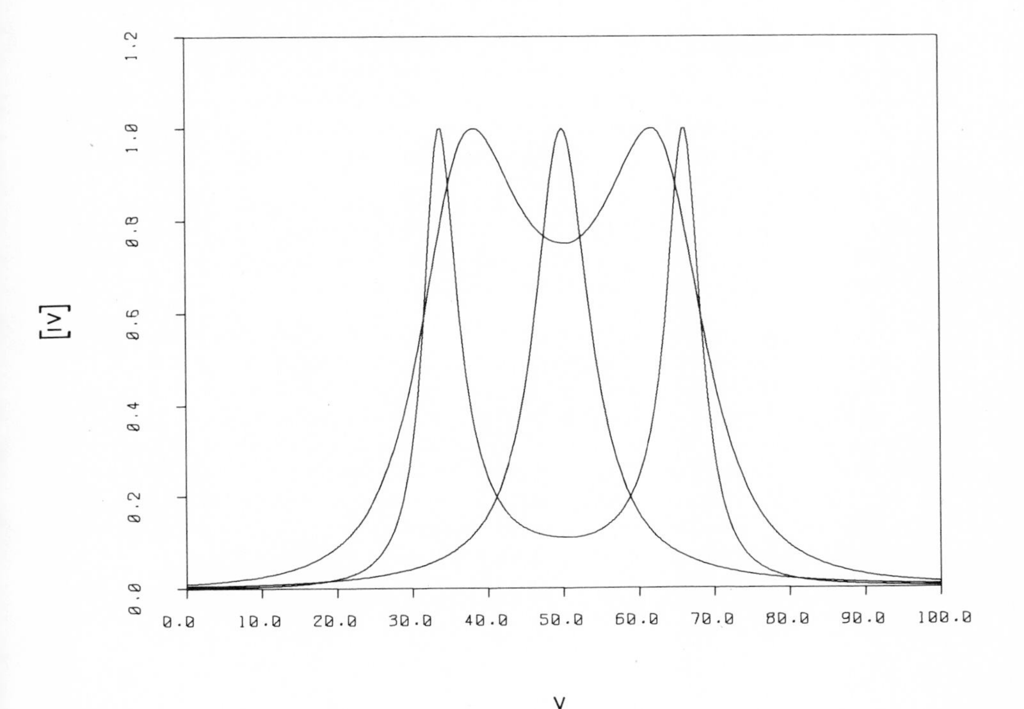

Fig. 25. Two-site NMR exchange.

$2\pi\tau(\nu_A - \nu_B) = \sqrt{2}$, and the singlet narrows as this value decreases (rate of exchange increases). In this example $0 \leq \nu \leq 100$, $\nu_A = 33.3$, $\nu_B = 66.7$, and τ is an input variable. The three curves are for $2\pi\tau(\nu_A - \nu_B) = 6$ (doublet), 2 (intermediate), and 0.55 (singlet).

Simulation programs of this sort are easily coded in the Culler-Fried language. And when the student has been using teletype, line printer, and Calcomp graphics, he is thrilled to sit at the keyboard, enter values for parameters, and find the curves displayed immediately on the scope.

During the fall term of 1970, a memo was sent to all students enrolled in physical chemistry courses. Interested students were invited to sign up for a special one-credit course meeting one evening each week. Eight students enrolled. After two hours the students were able to start programming the interactive graphics system. The only requirement was a documented project due at the end of the term. Most of the students chose their own projects, which included a simulation of van der Waals forces, a study of the solubility behavior of silver halides as a function of halide concentration, and an analysis of the Michaelis-Menten equation and Lineweaver-Burk plots of enzyme cooperativity. One of the students was so turned on by the system that he wanted more information about the NSF proposal. He was going to graduate school and wanted to get a terminal.

One of my graduate students has taken scores of 35-mm slides and hundreds of feet of Super-8 film directly from the scope. The objective is to develop several modules containing slides, prints, film loops, and additional information that may be of use to students and teachers.

As soon as the NSF grant terminated, a project to develop a similar interactive graphics system at Pitt was initiated. This project has three phases: developing an interpreter to perform the desired computations; interfacing the graphics terminal to Pitt's PDP-10; and implementing the software required to display curves on the scope and plotter and list programs and files on the teletype. The first phase of this project is essentially complete. PIGS (Pitt's Interactive Graphics System) is a Fortran IV program that interprets operators and operands coded in a subset of the Culler-Fried language. The output is currently teletype art.

Figure 19 is the output of a PIGS program. The overlapping radiodecay system is described by

$$A_{tot}(t) = A_{10} \exp(-0.693t/\tau_1) + A_{20} \exp(-0.693t/\tau_2).$$

The PIGS program uses the following variable names for scalars: $A = \tau_1$, $B = \tau_2$, $C = A_{10}$, $D = A_{20}$, $Q = \ln 2$, and $R = t_{max} = 4\tau_2$; and for vectors: $A = A_1(t)$, $B = A_2(t)$, $T = t$ $(0 < T < 4\tau_2)$ and $S = A_1(t) + A_2(t)$. A sample execution follows.

```
      .
      .
      .
.RUN PIGS

    ENTER INPUT LINE (72A1)

>LVL1 LOAD 60 STORE A LOAD 120 STORE B

    ENTER NEXT LINE

>LOAD 2000 STORE C LOAD 1000 STORE D MY RADIO

    USER PROGRAM RADIO CALLED FROM TTY

  X MIN = 0.000E - 01    X INC = 9.600E + 00    X MAX = 4.800E + 02
  Y MIN = 7.812E + 00    Y INC = 4.603E + 01    Y MAX = 3.000E + 03
```

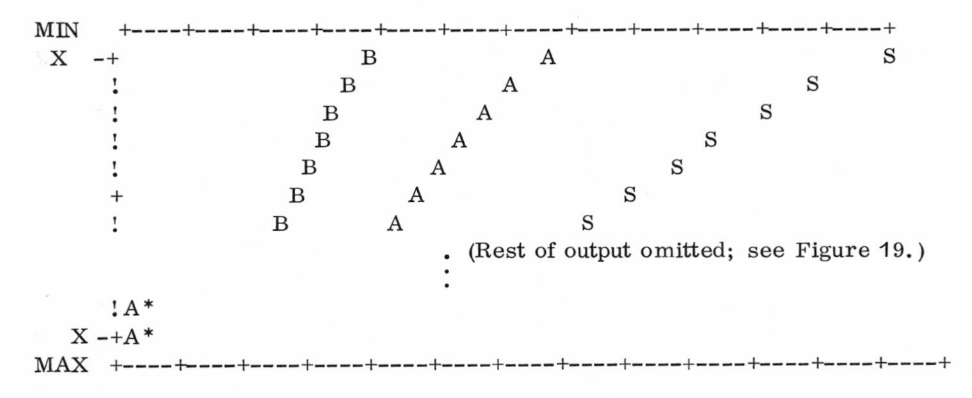

```
MIN    +----+----+----+----+----+----+----+----+----+----+----+----+
 X   -+                B                A                          S
      !              B              A                        S
      !            B              A                      S
      !           B             A                    S
      !          B            A                  S
      +         B           A                S
      !        B          A            S
                               .  (Rest of output omitted; see Figure 19.)
                               .
                               .

      !A*
   X -+A*
MAX  +----+----+----+----+----+----+----+----+----+----+----+----+----+
```

```
    END OF USER PROGRAM RADIO
    ENTER NEXT LINE

>$

    EXIT
    CPU TIME = 2.22 SEC
```

The command +LVLI" is a scalar mode declaration, i.e., all variables that follow are designated as scalers. The values 60, 120, etc., are assigned the indicated variable names. The command "MY" informs PIGS that the input is going to be transferred to a disk file whose name follows (here RADIO). A listing of the file RADIO is given:

```
LOAD 2 LOG STORE Q LOAD B * 4 STORE R
LVL2 CTX 51 COORD 0,R1 STORE T
LOAD T * Q1 / A1 NEG EXP * C1 STORE A
LOAD T * Q1 / B1 NEG EXP * D1 STORE B
LOAD A + B STORE S SUB T DISP A,B,S
$
```

The natural logarithm of 2 is stored in Q and $4\tau_2$ is computed and stored in R. Then the level of operation is switched to vector mode with the command "LVL2." The operator "CTX" sets the dimensionality of the vectors, here 51. The commands "COORD 0, R1 STORE T" fill the 51 components of T with the values (for R = 4*120 = 480), {0, 9.6, 19.2, ···460.8, 470.4, 480}. The next two lines compute A(t) and B(t). The scalar A (A_{10}) is differentiated from the vector A(t) by the 1. The sum of A(t) and B(t) is stored in S(t). The commands "SUB T DISP A, B, S" instruct PIGS to place T in the X register (abscissa) and then plot A, B, and S vs T.

The file RADIO was created and edited using a text editor. After execution of RADIO (the key symbol $ indicates the end of stored program mode), the terminal again becomes the principal input device, and PIGS prompts the user for more input. A user could easily vary the four parameters of this problem (τ_1, τ_2, A_{10}, A_{20}) until his objective was attained. Many of the simulations discussed in this chapter reduce to simple programs like RADIO.

PIGS is currently providing teletype art interactive graphics. The next step is to negotiate with a vendor and interface the scope and plotter to the PDP-10.

An interactive graphics package adds a totally new dimension to the use of computers to simulate chemical systems. Soon Pitt will provide its chemistry students with both a library of simulation programs on an interactive graphics system and an optional short course in PIGS programming. Then the students will be able to add this powerful appendage to their natural calculating abilities.

F. Discussion

This section of the chapter has described four approaches to computer-augmented learning: the optional PIL programming course; the library of simulation and data reduction programs; the numerical methods elective; and interactive graphics. The objective is to provide computing power as a tool so that those students who are interested in going beyond the textbook may do so unencumbered by arithmetic detail.

With a time-sharing system, an interpretive language like PIL or BASIC, and adequate access to the computer, the optional programming course is a most effective way to use computers in chemical education. The elements of computer programming can be easily covered by examples in a few lectures. Those students who wish to go further can and will get the information from programming manuals.

The simulation and data reduction programs can be implemented on a batch-processing, a remote-job entry, or a time-sharing system. Several of the programs are algorithmically trivial, involving no more than evaluation of a function with a few parameters over a range of the independent variable, and can be implemented using programmable desk calculators. The cost in execution time of these programs is quite low. Few require more than 1 or 2 seconds of CPU time. Also, several libraries of documented programs have been established [44].

The numerical methods course and interactive graphics require a more serious commitment of time and resources. The numerical methods course has been very gratifying. The students have developed mathematical and programming skills that will certainly be useful in graduate work. And several topics in chemistry were introduced that most of the Pitt undergraduates would not encounter until graduate school.

Graphics can be readily implemented using a teletype or line printer plot program. Incremental plotters are also readily available, and most include a library of plotting routines. An interactive graphics system requires a time-sharing system, a graphics terminal, and an interactive graphics language. The ease with which chemical systems can be simulated using an interactive graphics system certainly justifies the time and expense involved.

IV. THE COMPUTER-GENERATED REPEATABLE EXAMINATION SYSTEM (CGRE)

One of the most traumatic experiences for freshman chemistry students is the first hour exam. The CGRE system being developed here is designed to (among other things) minimize this trauma. To be first used during the Fall 1972 term, the CGRE system will give students the opportunity to take the first hour exam on a repeatable basis. Students will be allowed to take the exam up to five times during the first ten weeks of the term (the schedule is arbitrary). The tests will be administered by proctors in rooms reserved several hours a week for testing. The proctor will select one of several hundred computer-generated exams of equivalent difficulty, remove the answer key, and administer the test. When the student has finished he can help the proctor score the exam using the answer key. The score will then be recorded and the student will retain the graded test. The answer sheet, with missed items indicated, will become input to an item-analysis program.

Moore and Prosser at Indiana University have demonstrated the feasibility and advantages of a CGRE system [45]. They generate tests of equivalent difficulty from a database containing complete test items. The approach taken at Pitt is to write subroutines defining the item format. For example,

> HOW MANY GRAMS OF ___, ___, CAN BE MADE FROM ___ GRAMS OF ___ AMD ___ GRAMS OF ___?

The blanks are filled in by a subroutine that identifies this item, and contains lists of compounds and elements, ranges for random numbers, etc. For example,

> HOW MANY GRAMS OF ALUMINUM OXIDE, AL203, CAN BE MADE FROM 42.3 GRAMS OF AL AND 40.2 GRAMS OF O2?
>
> A) 130 B) 122
> C) 79.9 D) 85.4
> E) 82.5

Here the elements Al and O_2 were selected from a list of elements (34 possible cations and 11 possible anions). The amounts are chosen so that either reagent may be limiting. The subroutine finds the limiting reagent (Al in this case), computes the correct answer (C), computes the answer assuming the other reagent is limiting (D), and three other incorrect answers. These numbers and the order in which the options are printed are generated using random numbers generated over the appropriate ranges. The subroutine that generates this item requires approximately 0.15 second of CPU time (more than 90% of this is input-output time) and occupies 500 36-bit words of storage.

By September 1972 the test generating program will consist of over 100 Fortran subroutines. The primary usage of the CGRE system will be to generate the first hour exam for the general chemistry course. This exam covers four areas that are sometimes adequately treated in high school chemistry courses: stoichiometry, gas laws, solids, and nomenclature. A more detailed outline is indicated in Table 6.

TABLE 6

Topics Covered by the CGRE System

I. STOICHIOMETRY	II. GAS LAWS
A. Atoms and Molecules	A. General Gas Laws
B. Chemical Reactions	B. Dalton's and Graham's Laws
C. Redox Reactions	C. Kinetic Theory and Miscellaneous
D. Electrochemistry (Faraday's laws)	III. SOLIDS
E. Colligative Properties	A. Unit Cells
F. Heats of Reaction (Hess's law)	B. Density Calculations
G. Solution Stoichiometry	
H. Mixed Stoichiometry	IV. NOMENCLATURE
I. Laboratory Stoichiometry	

Within each topic, the items generated are classified as easy, of moderate difficulty, or difficult. The CGRE system allows the instructor to specify the number of items, the coverage, and the difficulty of the items in the test. The program prompts the instructor for the number of easy, moderate, and difficult items from topic IA, IB, etc. The program also has a number of special options, including lists of items to be included or excluded.

The current CGRE system uses the multiple-choice format exclusively. Later versions may include other formats, e.g., true false, fill-in, etc. Several representative examples follow.

1). A NEUTRAL ATOM WHICH HAS 63 ELECTRONS AND AN ATOMIC MASS OF 132 HAS:

A) A NUCLEUS CONTAINING 63 NEUTRONS
B) A NUCLEUS CONTAINING 69 NEUTRONS
C) A NUCLEUS CONTAINING 69 PROTONS
D) NO OTHER ISOTOPES
E) AN ATOMIC NUMBER OF 69

2). A HYPOTHETICAL ELEMENT HAS THE FOLLOWING ISOTOPIC COMPOSITION:

ISOTOPE	MASS	PERCENT ABUNDANCE
70	69.9713	55.5
72	71.9256	44.5

WHAT IS THE ATOMIC WEIGHT OF THE ELEMENT?

A) 55.5*69.9713 + 44.5*71.9256
B) (55.5*69.9713 + 44.5*71.9256) / 100
C) 69.9713 + 71.9256
D) (69.9713 + 71.9256) / 2
E) NONE OF THE ABOVE

3). WHICH ONE OF THE FOLLOWING MEASUREMENTS OR TECHNIQUES CAN BE USED TO FIND THE ATOMIC OR MOLECULAR WEIGHT OF A SUBSTANCE?

A) MASS SPECTROSCOPY
B) DENSITY OF LIQUID STATE
C) THERMAL CONDUCTIVITY
D) BOILING POINT OF PURE LIQUID
E) CHEMICAL ANALYSIS

4). CONSIDER THE FOLLOWING HYPOTHETICAL UNBALANCED REACTION:

Ti + S + 02 ---> Ti(SO4)2

GIVEN: .430 GRAM-ATOMS OF Ti, 29.2 GRAMS OF S, AND 28.3 LITERS OF O2 AT STP, HOW MANY MOLES OF Ti(SO4)2 CAN BE PREPARED?

A) 1/.430
B) (28.3 / 22.4) / 4
C) (29.2 / 32.1) / 2
D) .430
E) NONE OF THE ABOVE

5). A SAMPLE OF O2 WAS COLLECTED OVER WATER AT 42.4 DEGREES C, WHEN THE BAROMETRIC PRESSURE WAS 726 MM OF HG. THE VOLUME OF THE GAS AS IT WAS COLLECTED WAS 3.05 LITERS. WHAT WOULD THE VOLUME OF DRY O2 BE AT STP? (THE VAPOR PRESSURE OF H2O AT 42.4 DEGREES IS 63.5 MM OF HG.)

A) 1.71 LITERS
B) 4.05 LITERS
C) 3.69 LITERS
D) 2.30 LITERS
E) NONE OF THE ABOVE

6). AMMONIA IS PREPARED ACCORDING TO THE FOLLOWING REACTION IN THE GAS PHASE:

N2 + 3H2 ---> 2NH3

IF THE REACTION CONDITIONS ARE MAINTAINED AT STP, WHICH OF THE FOLLOWING STATEMENTS IS INCORRECT?

A) 22.4 LITERS OF N2 WILL REACT WITH 3 * 22.4 LITERS OF H2 TO FORM 2 * 22.4 LITERS OF NH3
B) 22.4 LITERS OF N2 WILL REACT WITH 6.0 GRAMS OF H2 TO FORM 34 GRAMS OF NH3.
C) 129 LITERS OF NH3 CAN BE PREPARED FROM 258 LITERS OF N2 GIVEN AN EXCESS SUPPLY OF H2.
D) 28.0 GRAMS OF N2 WILL REACT WITH 3 * 22.4 LITERS OF H2 TO FORM 34.0 GRAMS OF NH3.
E) 129 LITERS OF H2 WILL REACT WITH 42.9 LITERS OF N2 TO FORM 85.8 LITERS OF NH3.

7). WHAT WEIGHT OF ETHYLENE GLYCOL (HOCH2CH2OH, MW = 62.1 G/MOLE) MUST BE ADDED TO 5000 GRAMS OF WATER TO PRODUCE AN ANTIFREEZE SOLUTION THAT WOULD PROTECT A CAR RADIATOR DOWN TO 0.2 DEGREES FARENHEIT (= -17.7 DEGREES CENTIGRADE)? (ETHYLENE GLYCOL IS A NONELECTROLYTE.)

A) (6.21 * -1.86) (5 * -17.7)
B) (62.1 * -17.7 * 5) / (-1.86)
C) (62.1 * -1.86 * 5) / (0.2)
D) (62.1 * -17.7) / (5 * -1.86)
E) NONE OF THE ABOVE

8). SUPPOSE 0.10 FARADAY OF CHARGE IS PASSED THROUGH 500 ML OF 1.00 MOLAR NaCl SOLUTION. THE ELECTRODE REACTIONS ARE:

ANODE 2Cl(-) ---> Cl2 + 2E(-)
CATHODE 2H2O + 2E(-) ---> 2OH(-) + H2

ONLY ONE STATEMENT BELOW IS TRUE FOR THIS REACTION. WHICH IS THE CORRECT STATEMENT?

A) 8.0 G OF NAOH COULD BE RECOVERED FROM THE SOLUTION.
B) 1.12 L OF Cl2 ARE PRODUCED (STP)
C) THE CONCENTRATION OF Cl(-) IN THE SOLUTION IS 0.90 MOLAR.
D) THE SOLUTION CONTAINS 0.40 MOLES OF Na(+).
E) 0.20 MOLES OF OH(-) ARE PRODUCED.

9). THE SAME QUANTITY OF ELECTRICITY THAT CAUSED THE DEPOSITION OF 181 GRAMS OF SILVER FROM A AGNO3 SOLUTION WAS PASSED THROUGH A SOLUTION OF A TELLURIUM SALT CONTAINING TE CATIONS OF UNKNOWN CHARGE. 215 GRAMS OF TE WERE DEPOSITED. WHAT IS THE CHARGE OF THE CATIONS?

A) 3 B) 2
C) 1 D) 4
E) NONE OF THE ABOVE

10). SELECT THE PROPER SET OF NAMES FOR THE FOLLOWING FOUR COMPOUNDS:

1. PT(C2O4)2. 2. CU3N2. 3. MG3(PO4)2. 4. TI(CRO4)2.

A) 1. PLATINUM(IV) OXALATE 2. COPPER(II) NITRIDE
3. MAGNESIUM PHOSPHATE 4. TITANIUM(IV) DICHROMATE

B) 1. PLATINUM(IV) OXALATE 2. CURIUM(II) NITRIDE
3. MAGNESIUM PHOSPHATE 4. TITANIUM(IV) CHROMATE

C) 1. PLATINUM(IV) OXALATE 2. COPPER(II) NITRIDE
3. MAGNESIUM PHOSPHATE 4. TITANIUM(IV) CHROMATE

D) 1. PLATINUM(II) OXALATE 2. CURIUM(II) NITRIDE
3. MAGNESIUM PHOSPHATE 4. TITANIUM(IV) CHROMATE

E) NONE OF THE ABOVE

Items 1, 2, and 10 are considered easy, items 5, 6, and 7 are considered of moderate difficulty, and items 3, 4, 8, and 9 are considered difficult.

This project will affect the freshman chemistry course in several favorable ways. The traumatic first hour exam will be given on a repeatable basis. The "bad day" excuse will disappear. The examination room will be relatively small, and therefore the nervousness and tension that pervades amphitheater testing sections will be minimized. The make-up exam problem will be eliminated. The test will hopefully allow the lecturer to spend less time on stoichiometry and gas laws and more time on atomic structure, bonding, periodic properties, descriptive chemistry, etc. The system can be "tuned" via an item analysis system so that eventually reliable examinations of arbitrary coverage and difficulty can be obtained on demand. The database of item formats can be used to generate entire exams, parts of exams, tutorial quizzes, make-up exams, etc. It can also be used in a tutorial mode for student self-study. The entire database, cross-referenced with the textbook, a programmed textbook, a problem book, and the CAI lessons, will be made available to students.

The exam will also serve a diagnostic purpose. Students may take the exam during the first few days of the term and decide to drop the course if this is warranted. The bright students will relatively quickly get past this hurdle and concentrate their efforts on the more intellectually stimulating topics in freshman chemistry (atomic theory, bonding, molecular structure, etc.).

V. CONCLUSIONS

This chapter has covered three applications of computers in chemical education. Two of these are tailored to the freshman chemistry curriculum: computer-assisted instruction and the computer-generated repeatable examination system. The third category, computer-augmented learning, is appropriate for all levels of education.

CAI is clearly the most expensive of the three approaches. There are many requirements: a reliable and responsive time-sharing system, an adequate supply of terminals, an appropriate programming language, adequate computer storage, a lesson designer, programmers, etc. The programmers who have worked with me have been chemistry graduate and undergraduate students. Including line-by-line editing by me and one of my colleagues (Dr. L. M. Epstein), it has taken from 100 to 200 hours to design, code, debug, and edit one hour of interactive CAI. At $2.00 per hour this comes to between $200 and $400 per hour in programmer time alone. This might be a one-time expense. However, the CAI programs discussed here have undergone continuous revision during the past two years. They are currently being rewritten according to the flowchart in Fig. 3.

The 26 lessons described in Table 1 require approximately 50,000 36-bit words of disk storage. To run CATALYST, PIL, and one of the CAI lessons, the student requires approximately 21,000 36-bit words of core. The execution time is approximately 20 cpu seconds per hour of CAI. At $180 per CPU-hour, this component is $1.00 per CAI-hour. Other costs include rental or purchase of terminals and acoustic couplers, telephone, terminal connect time, the cost to store the CAI lessons for easy access, and supplies.

Do the advantages of CAI justify this expenditure of time and resources? I suggest that it is too early to attempt a judgement on the cost effectiveness of CAI. I am encouraged by the enthusiasm of my students for the system, received in oral and written form (see above). I suspect that the usage is greater among the upper 20% of the class than it is with the below average students for whom the system has been designed. However, the lessons are being used by an increasing percentage of the students. And I fully expect that over the years we will develop a CAI system that will provide effective tutoring to those students who avail themselves of it.

The computer-generated repeatable examination system is being developed at the time of this writing. Several 20-item exams have been generated from 100 question formats. The subroutines containing the code to generate variable items require on the average approximately 1000 36-bit words, 0.25 second of CPU time, and are clearly I/O bound. Therefore, it takes approximately five seconds of CPU time to generate one exam. I expect to reduce the I/O time substantially before going into production during the Fall 1972. It is my hope that this system will minimize the trauma associated with the first freshman chemistry exam and will allow the lecturer to spend less time reviewing topics already covered in high school chemistry courses.

It is with what I have called computer-augmented learning that the return on an instructor's investment of time and energy is greatest. By teaching programming languages and/or making available a set of tools with which students can solve some interesting and pedagogically useful problems, we can significantly affect chemical education. For example, during the winter term of 1972, I made several optional assignments in my section of a sophomore-level quantitative analysis course. Some typical examples follow:

1. Calculate the solubility of a series of sparingly soluble salts as a function of pH, for example, AgOAc, Ag_2CO_3, and Ag_3PO_4. Plot the results on a single graph and discuss the results.

2. Calculate the solubility of a series of sparingly soluble salts in aqueous ammonia as a function of $\log[NH_3]$. Plot and discuss.

3. Calculate the solubility of a series of sparingly soluble salts as a function of excess anion. Determine the anion concentration for which the solubility is a minimum. Plot and discuss.

4. Plot pH vs log Ca for several monoprotic weak acids HA. Show that as Ca approaches zero, pH approaches 7. Indicate the failure of the first and second approximations to the complete solution.

5. Calculate several weak acid-strong base titration curves as a function of Ka and Ca. Plot and discuss.

6. Calculate several diprotic or triprotic acid-strong base titration curves. Plot and discuss.

7. Plot the fraction of the four phosphate species and the five EDTA species as a function of pH.

8. Plot pH as a function of log C for a series of buffer systems ($C = C_{HA} = C_A$).

9. Simulate M^{2+}-EDTA titration curves as a function of K_f, C_m, and pH. Plot and discuss.

10. Make an EDTA titration feasibility plot. Plot $\Delta pM_{101\%}$-$pM_{99\%}$ vs pH as a function of C_m and K_f. Plot and discuss.

11. Calculate the potential-pH behavior of several $M^{z+}|M$ electrodes in aqueous ammonia solutions. Plot and discuss.

12. Calculate a series of potentiometric titration curves. Plot and discuss.

The Fortran IV programs SOL, HA, EDTA, and NERNST were made available for problems 1, 5, 10, and 11, respectively. Interested students were encouraged to attend my four-hour optional PIL course. With this background most of them could readily write their own simulation programs.

The students were told that completion of these optional assignments would contribute toward their grade only if they were borderline cases. Of the 30 students in the class, 12 completed at least one of the 16 optional assignments and five students completed five or more assignments. Most of these students used the Fortran programs and wrote short PIL programs to generate the data which they then hand-plotted. Some students used their slide rules to make the required calculations.

The student feedback on the use of the computer as a tool to solve some interesting and illustrative problems was very enthusiastic. I highly recommend the use of optional assignments to challenge the better students in the class.

A key word in this discussion is the word "optional." I do not believe that the computer applications described here should be required of all students in any course. The demand for computer terminals usually far exceeds the supply, especially during prime time. The students must learn some of the system command language in order to log on, load and execute the desired programs, and log off. Most computer systems still have rather rigid formating specifications, are rather intolerant of mistakes, and occasionally, the time-sharing system crashes. Some students are temperamentally unable to tolerate these idiosyncrasies and should be spared.

A two-term experiment in teaching programming was recently conducted in the junior-level physical chemistry laboratory course. The students were given two hours of PIL programming instruction and were then expected to write a linear regression program. Several of the students had no trouble with this assignment. However, for many others it was too much too fast. I think computer opportunities should be optionally provided at all levels, from CAI to interactive graphics.

I feel that computers can definitely add a new dimension to chemical education. With care, we can provide tools with which our students can both more easily grasp the traditional chemical concepts and also explore phenomena that are beyond their grasp using conventional techniques.

ACKNOWLEDGMENTS

Support of this research from two National Science Foundation grants is gratefully acknowledged. Grant GJ-686 supported the computer network providing the Culler-Fried interactive graphics system (David O. Harris, University of California, Santa Barbara, principal investigator). Grant GU-3184, a Science Development Grant to the University of Pittsburgh, provided salary support for the author and several programmers, rental of terminals, supplies, etc.

Several man-years of software development have been described in this chapter. The following are the major contributors of the CAI material: E. Conrad, D. Doerfler, W. Jordan, and E. Tratras. Most of the Fortran IV programs were written in conjunction with J. Cassidy, D. Doerfler, D. Hawkins, S. Levitt, and B. Swisshelm. S. Levitt wrote the programs for the Culler-Fried graphics system. S. Liberman and W. Sliwinski were involved with design and coding of PIGS. The following programmers have been the major contributors to the CGRE system: M. Felice, D. Hawkins, M. Juves, W. Sliwinski, and R. Smith.

Several of my colleagues have assisted me in this work. Dr. L. M. Epstein wrote items for the CGRE system and contributed many hours of careful editing of both CAI lessons and CGRE items. Dr. J. F. Coetzee suggested the EDTA and TITR programs, Dr. F. Kaufman suggested BOX, and Dr. D. W. Pratt suggested ENTROPY.

The Computing Center has cooperated fully with all aspects of this work. The administration should be commended for providing without charge extensive computing facilities for undergraduate education. The assistance of the Computer Center Staff on numerous occasions is also gratefully acknowledged.

REFERENCES

1. G. Badger, "PIL/L: Pitt Interpretive Language for the IBM System/360 Model 50," Univ. of Pittsburgh, rev. Feb., 1969.

2. (a) R. Blanc, "CATALYST: A Computer-Assisted Teaching and Learning System for a General Purpose Time-Sharing System," M.S. Thesis, Univ. of Pittsburgh, 1968. (b) T. Dwyer, "A Lesson Designer's Guide

to CATALYST and the CATALYST/PIL Interface," Department of Computer Science, Univ. of Pittsburgh, 1969.

3. (a) E. W. Ewig, J. T. Gerig, and D. O. Harris, J. Chem. Ed., 47, 97 (1970). (b) B. D. Fried, On-Line Computing (W. J. Karplus, ed.), McGraw-Hill, New York, 1967, pp. 131-178.

4. (a) R. C. Grandey, J. Chem. Ed., 48, 791 (1971). (b) S. G. Smith, J. Chem. Ed., 47, 608 (1970). (c) S. Castlebury and J. J. Lagowski, J. Chem. Ed., 47, 91 (1970).

5. T. Dwyer, Comm. ACM, 15, 21 (1972).

6. J. Ott, R. Goates, and H. Hall, J. Chem. Ed., 48, 515 (1971).

7. (a) Cf. the subject index in the December issues of J. Chem. Ed. under "Computer Use." (b) Proceedings of the Conference on Computers in Chemical Education and Research, DeKalb, Illinois, July, 1971. (c) Proceedings of the Conference on Computers in the Undergraduate Curricula: I, University of Iowa, Iowa City, 1970; II, Dartmouth College, Hanover, New Hampshire, 1971; III, Southern Regional Education Board, Atlanta, Georgia, 1972.

8. B. Carnahan, H. A. Luther, and J. O. Wilkes, Applied Numerical Methods, Wiley, New York, 1969, p. 171.

9. W. J. Blaedel and V. W. Meloche, Elementary Quantitative Analysis: Theory and Practice, Harper and Row, New York, 1963.

10. C. J. Nyman and R. E. Hamm, Chemical Equilibrium, Raytheon, 1968, p. 28ff.

11. B. Carnahan et al., Applied Numerical Methods, Wiley, New York, 1969, p. 178.

12. W. Kauzmann, Quantum Chemistry, Academic, New York, 1957, p. 188ff.

13. (a) E. A. Moelwyn-Hughes, Physical Chemistry, Chaps. 8-12, Pergamon, London and New York, 1961. (b) N. B. Colthrup et al., Infrared and Raman Spectroscopy, Academic, New York, 1964, p. 463ff.

14. J. O. Hirschfelder, J. Chem. Phys., 8, 431 (1940).

15. (a) J. A. Pople, W. G. Schneider, and H. J. Bernstein, High-Resolution Nuclear Magnetic Resonance, Chap. 6, McGraw-Hill, New York, 1959. (b) J. W. Emsley, J. Feeney, and L. H. Sutcliffe, High Resolution Nuclear Magnetic Resonance Spectroscopy, Vol. 1, Chap. 8, Pergamon, London and New York, 1965. (c) K. B. Wiberg and B. J. Nist, The Interpretation of NMR Spectra, Benjamin, New York, 1962.

16. K. B. Wiberg, Computer Programming for Chemists, Benjamin, New York, 1965, p. 189ff.

17. (a) A Streitwieser, Molecular Orbital Theory for Organic Chemists, Wiley, New York, 1961. (b) K. B. Wiberg, Physical Organic Chemistry, Wiley, New York, 1964.

18. B. Carnahan et al., Applied Numerical Methods, Chap 5, Wiley, New York, 1969.

19. (a) B. Carnahan et al., Applied Numerical Methods, Chap 8, Wiley, New York, 1969. (b) R. H. Moore and R. K. Ziegler, "The Solution to the General Least Squares Problem with Special Reference to High-Speed Computers," Los Alamos Reports, LA 2367 (1960). (c) W. F. Wentworth, J. Chem. Ed., 42, 96, 162 (1965). (d) W. J. Youden, Statistical Methods for Chemists, Wiley, New York, 1951, p. 41ff.

20. (a) B. Carnahan et al., Applied Numerical Methods, Wiley, New York, 1969, p. 319ff. (b) J. W. Swinnerton and W. W. Miller, J. Chem. Ed., 36, 485 (1959).

21. (a) B. Carnahan et al., Applied Numerical Methods, Wiley, New York, 1969, p. 363ff. (b) M. J. Romanelli, Mathematical Methods for Digital Computers (A. Ralston and H. S. Wilf, eds.), Vol. 1, Chap. 9, Wiley, New York, 1960.

22. T. R. Crossley and M. A. Slifkin, Progress in Reaction Kinetics (G. Porter, ed.), Vol. 5, Pergamon, London and New York, p. 409ff.

23. G. L. Hemphill and J. M. White, J. Chem. Ed., 49, 121 (1972).

24. (a) B. Carnahan et al., Applied Numerical Methods, Wiley, New York, 1969, p. 250ff. (b) J. Greenstadt, Mathematical Methods for Digital Computers (A. Ralston and H. S. Wilf, eds.), Vol. I, Chap. 7, Wiley, New York, 1960.

25. K. J. Johnson, J. Chem. Ed., 47, 819 (1970).

26. K. J. Johnson, Numerical Methods in Chemistry, 1971. (This is the first draft of the text for the numerical methods course. I have also used T. R. Dickson, The Computer and Chemistry, Freeman, 1968.)

27. The WATFOR compiler provides rapid Fortran IV compilation with extensive error diagnostics. Cf. P. Cress, P. Dirksen, and J. W. Graham, Fortran IV with WATFOR, Prentice-Hall, Englewood Cliffs, New Jersey, 1968.

28. C. Froberg, Introduction to Numerical Analysis, Addison-Wesley, Reading, Massachusetts, 1965, p. 73.

29. R. S. Varga, Matrix Iterative Analysis, Prentice-Hall, Englewood Cliffs, New Jersey, 1962, p. 56ff.

30. (a) C. Froberg, Introduction to Numerical Analysis, Chap. 4, Addison-Wesley, Reading, Massachusetts, 1965. (b) B. Carnahan et al., Applied

Numerical Methods, Chap. 5, Wiley, New York, 1969.

31. (a) L. G. Kelly, Handbook of Numerical Methods and Applications, Addison-Wesley, Reading, Massachusetts, 1967, p. 68ff. (b) B. Carnahan et al., Applied Numerical Methods, Wiley, New York, 1969, pp. 574-575.

32. A. Savitsky and M. J. E. Golay, Anal. Chem., 36, 1627 (1964).

33. A. A. Frost and R. G. Pearson, Kinetics and Mechanism, Wiley, New York, 1961, p. 173ff.

34. A. G. Desai, H. W. Dodgen, and J. P. Hunt, J. Am. Chem. Soc., 91, 5001 (1969).

35. (a) B. Carnahan et al., Applied Numerical Methods, Chap. 2, Wiley, New York, 1969. (b) C. Froberg, Introduction to Numerical Analysis, Chap. 10, Addison-Wesley, Reading, Massachusetts, 1965. (c) T. R. Dickson, The Computer and Chemistry, Chap. 8, Freeman, New York, 1968.

36. (a) B. Carnahan et al., Applied Numerical Methods, Chap. 6, Wiley, New York, 1969. (b) C. Froberg, Introduction to Numerical Analysis, Chap. 14, Addison-Wesley, Reading, Massachusetts, 1965.

37. J. Ortega, Mathematical Methods for Digital Computers (A. Ralston and H. S. Wilf, eds.), Vol. 2, Chap. 4, Wiley, New York, 1967.

38. A. A. Bothner-By, Computer Programs for Chemistry (D. F. Detar, ed.), Vol. 1, Chap. 3, Benjamin, New York, 1968.

39. J. A. Pople and D. L. Beveridge, Approximate Molecular Orbital Theory, McGraw-Hill, New York, 1970.

40. (a) E. A. Ogryzlo and G. Porter, J. Chem. Ed., 40, 258 (1963). (b) W. T. Bordass and J. W. Linnett, J. Chem. Ed., 47, 672 (1970). (c) M. J. S. Dewar and J. Kelener, J. Chem. Ed., 48, 496 (1971). (d) A. C. Wahl, Science, 151, 961 (1966). (e) T. H. Dunning and N. W. Winter, J. Chem. Phys., 55, 3360 (1971). (f) W. England, L. S. Salmon and K. Ruedenberg, Topics in Current Chemistry, Vol. 23, Springer-Verlag, Berlin, 1971, pp. 31-123.

41. J. C. Davis, Jr., Advanced Physical Chemistry, Chap. 8, Ronald, 1965.

42. H. Williamson, Comm. ACM, 15, 100 (1972); cf. also Ref. 40b and 40d.

43. J. A. Pople, W. G. Schneider, and H. J. Bernstein, High Resolution Nuclear Magnetic Resonance Spectroscopy, Vol. 1, Chap. 10, Pergamon, London and New York, 1965.

44. For lists of chemistry computer programs available the reader should write to the author, to Dr. R. C. Collins, Department of Chemistry, Eastern Michigan University, Ypsilanti, Michigan 48197, or to Dr. J. R. Denk, North Carolina Educational Computing Service, Research Triangle Park, North Carolina 27709.

45. F. Prosser and J. Moore, Proceedings of the Conference on Computers in Chemical Education and Research, DeKalb, Illinois, July, 1971, pp. 9-26.

Chapter 3

ANALOG AND HYBRID COMPUTATION

Frederick D. Tabbutt

The Evergreen State College
Olympia, Washington

I. INTRODUCTION

Hybrid computation is now within the reach of any institution that is considering a computer system. This is a recent development, for in the past, hybrid systems have been vast, complicated, and extremely expensive systems which could be afforded by only a few. With the arrival of high-speed, inexpensive, medium sized analog/hybrid systems and inexpensive minicomputers the systems costs have dropped remarkably. Now with the introduction of automatic patching it appears that we are on the threshold of a renaissance in hybrid computation.

But why use hybrid techniques? Why not stick to the familiar digital techniques? The advantage is that in dealing with computations or simulations involving the dynamics of natural systems the analog computer is fast — orders of magnitude faster than the fastest digital. The hybrid system supplements this speed with the memory and decision-making capability of the digital computer. With proper design a hybrid system combines the best of both computers.

There are many powerful techniques in analog programming which a hybrid system can exploit. Though familiar to electrical engineers, they are new to most chemists. This chapter will introduce these techniques and demonstrate them with examples from chemical and physical systems. The software structure of a hybrid system will be introduced, and finally the implications of current autopatching prototype systems will be explored.

A. Analog vs Digital

It is illuminating to examine the strengths and weaknesses of analog and digital computers and note the degree to which they are complementary.

TABLE 1

A Comparison of Analog and Digital Computers

	Analog	Digital
Operation	Parallel — computation time independent of problem complexity	Serial — computation time dependent on problem complexity
Speed (see below for detailed comparison)	Fast for differential equations, awkward for linear algebra	Slow for differential equations, fast for linear algebra
Memory	Limited to 140 to 700 bits (number of integrators × 14 bits, assuming electronic hold)	Enormous, can virtually always be made adequate
Precision	Adequate — 0.01% for low speed 0.1% for high speed	As precise as needed

TABLE 1 (continued)

	Analog	Digital
Accuracy	0.1%	For differential equations, not as good as precision because of integration approximations
Decision-making ability	Crude — based on modest complement of logic elements	Sophisticated — limited only by programmer's imagination

There has been considerable discussion, some of it heated, comparing the speed of analog and digital computers. The comparison is complicated by the fact that analog computers solve equations in a parallel manner so that the computation time is the same regardless of the complexity of the problem, while the serial operation of a digital computer makes it problem-dependent. Clancy [1] offered an interesting approach to the problem by specifying an average medium-sized analog computer, assuming that all components were

TABLE 2

Digital Equivalency of a Medium-Sized Analog Computer[a]

Component	Quantity in average medium-sized computer	Digital computer instructions req'd to simulate one component	Total no. of digital operations for analog component
Integrator	30	25 (4 point predictor)	750
Summer (3 input)	40	6	240
Inverter	10	2	20
Multiplier	64	3	192
Resolver	25	55	1375
Function generator	10	20 (table look-up)	200
Coefficient potentiometer	100	3	300
Total			3077

[a]Obtained from 55 IBM 7090 programs involving 250,000,000 instructions. The raw data were obtained from 30 FORTRAN source programs, five machine language programs, ten assemblies, and ten compilations.

operating at once and determining the time required for a digital computer to emulate these operations. (The composition of the average analog computer he used is given in Table 2.) The number of digital instructions required for one time step to implement each of the above analog operations is calculated as shown in Table 2. The total, 3077, is the number of digital instructions required to increment the digital computer a single Δt and simulate the capability of the typical medium-sized analog computer. To attach a specific time to execute the 3077 instructions requires one more approximation because the instructions vary in the time they require for execution. accordingly, an average scientific instruction time was arrived at through a Gibson mix, shown in Table 3 [1].

TABLE 3

Gibson Mix[a]

Operation	% instruction type in Gibson Mix
load and store	31.2%
add and subtract	6.1
multiply	0.6
divide	0.2
floating multiply	3.8
floating divide	1.5
floating add and subtract	6.9
search or compare	3.8
test and branch	16.6
shift	4.4
logical	1.6
no memory reference	5.3
indexing	18.0
Total	100.0%

[a]Obtained from 55 IBM 7090 programs involving 250,000,000 instructions. The raw data were obtained from 30 FORTRAN source programs, five machine language programs, ten assemblies, and ten compilations.

TABLE 4

Digital Computer Bandwidth

Digital computer	Gibson mix average execution time/operation (μ sec)	Execution time for 3077 operations (msec)	Bandwidth per Δt (cps)	Maximum problem frequency with 100 steps (solutions/sec)	Maximum problem frequency with 1000 steps (solutions/sec)
PDP-6	4.17	12.83	77.9	0.8	.08
DDP-224	4.06	12.49	80.1	0.8	.08
CDC-3200	3.70	11.38	87.9	0.9	.09
IBM 70941	3.23	9.94	100.6	1.0	.10
IBM 7094II	2.32	7.14	140.0	1.4	.14
GE 635	2.07	6.37	156.9	1.6	.16
UNIVAC 1108	1.66	5.11	198.0	2.0	.20
IBM 360/67	1.49	4.58	218.3	2.2	.22
CDC 6400	1.16	3.57	280.1	2.8	.29
IBM 360/75	0.75	2.31	432.9	4.3	.43
CDC 6600	0.29	0.89	1123	11.2	1.12

Clancy then multiplied the time for each line entry in Table 3 by the percent occurrence for it to obtain an average execution time. The Gibson mix average (GMA) for several computers is shown in Table 4. Note that the Gibson mix will always be greater than the cycle time. Multiplying the GMA by 3077 gives the time required for each of the digital computers to emulate one time step of the medium-sized analog. The reciprocal of this gives the bandwidth or number of computations per second. The analog-digital comparison requires one more step. So far only a single time step per computation has been considered. This is not adequate. Typically, 100 steps would be considered minimal, while 1000 would yield a solution that would appear to be continuous on an oscilloscope or an 8" x 11" sheet of chart paper and have analog accuracy over this range. It can be seen that for a high fidelity plot of 1000 points even the fastest computer tabulated will barely do one solution in a second. The present day analog computers are capable of a computation bandwidth of 1000 computations per second to an accuracy of 0.1% (which is comparable to the 4 point predictor used by Clancy in Table 2). Furthermore, they can display a variety of variables simultaneously at the same speed with a resolution equivalent to at least 1000 points/computation. Display time for the digital computer would have to be added to the strictly computational times tabulated in Table 4.

It is this breakneck computational speed coupled with display versatility which has made the analog computer a powerful interactive tool in real-time or faster-than-real-time simulations. Now that medium-sized analog computers are available for $50,000, a minicomputer for $5,000, and an interface for $5,000, fast, interactive hybrid systems are worthy of consideration for simulation studies.

II. METHODS OF PROGRAMMING ANALOG COMPUTERS

The programmer of an analog computer need no longer be an expert or even familiar with electronics or electrical circuitry. In contrast to the earlier analog systems, whose programmer had to be able to patch capacitors and resistors, present day systems have these elements permanently built into the machine. Consequently, a programmer need only consider the operation performed by each of the components. In other words, the relationship between output and input and the symbol to represent that operator are all that are required to perform analog programming. Writing the program, then, consists of sketching the appropriate components and showing how they are to be connected to each other.

Appendix I is a description of common analog and logic elements and the uniform graphics which have been agreed upon by a Simulation Councils, Inc. Committee. Simulation Councils (P.O. Box 2228, LaJolla, Calif.) is an organization specializing in analog and digital simulation techniques.

A number of sources [3-5] exist for the reader who wants to explore the hardware interior of one of these components. Moreover, there exist a number of introductions [3, 5-9, 28] to analog computation. This chapter will assume a familiarity with the symbols in Appendix I.

The ability to perform instantaneous integrations is the most important feature of an analog computer. In fact, an alias for the analog computer is electronic differential analyzer. Long experience has shown that nature is most effectively modeled using differential equations, both ordinary and partial. Therefore, from a natural scientist's point of view differential equations and the ability to solve them are very important mathematical tools. Consequently, the following discussion will deal with the analog computer only as a differential equation solver. Techniques for ordinary and partial differential equations will be presented.

A. Ordinary Differential Equation

An ordinary differential equation can be expressed in general form as

$$a_n \frac{d^n y}{dx^n} + a_{n-1} \frac{d^{n-1} y}{dx^{n-1}} \cdots + a_1 \frac{dy}{dx} + a_o y = f(x). \tag{1}$$

The distinguishing feature of an ordinary differential equation is that there is one independent variable, x. The coefficients can be constant, e.g.,

$$\frac{dN}{dt} + kN = 0, \quad \text{first order decay,} \tag{2}$$

and

$$\frac{d^2\psi}{dx^2} + \epsilon\psi = 0, \quad \text{Schrödinger equation for a free particle,} \tag{3}$$

or variable

$$(1 - x^2)\frac{d^2\theta}{dx^2} - 2x\frac{d\theta}{dx} + n(n + 1)\theta = 0, \quad \text{Legendre's differential equation, } \theta \text{ equation for the hydrogen atom.} \tag{4}$$

These equations are defined as linear ordinary differential equations (LODE), since the coefficient a_i is either a constant or a function of the independent variable. These are to be distinguished from the nonlinear ordinary differential equations (NODE). Examples of NODEs are

$$\frac{dC}{dt} + kC^2 = 0, \quad \text{a second-order reaction,} \tag{5}$$

and

$$\frac{d^2\theta}{dt^2} + \frac{g}{r} \sin \theta = 0, \qquad \text{motion of a pendulum.} \tag{6}$$

Though the difference between NODEs and LODEs is significant in terms of the difficulty of an analytical solution, the analog program is similar for both.

1. General Method

The most powerful and at the same time simple approach to the analog solution of ODEs is the general method. The general method was discovered nearly 100 years ago on a mechanical analog computer by Lord Kelvin [12]. The method will be demonstrated using the differential expression for the kinetics of a parallel first- and second-order decay as an example:

$$\frac{dC}{dt} + k_1C + k_2C^2 = 0. \tag{7}$$

(1) Rearrange the differential equation so that the highest-order derivative stands alone with a coefficient of unity and is equal to all other terms in the equation.

$$-\frac{dC}{dt} = k_1C + k_2C^2. \tag{8}$$

(2) Assume that the highest order derivative is available as an input to an integrator.

$$\frac{dC}{dt} \quad \text{is available.}$$

(3) Connect in series as many integrators as are required to produce the lower order derivative terms. Since this is a first-order differential equation only one integrator is required:

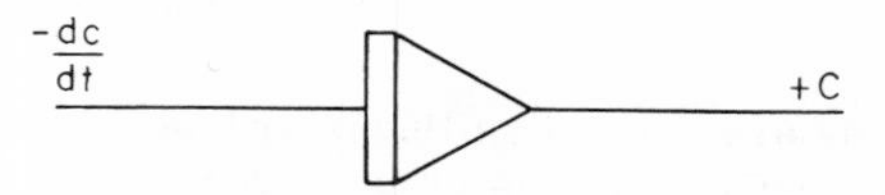

Then to produce the k_1C and k_2C^2 terms we have:

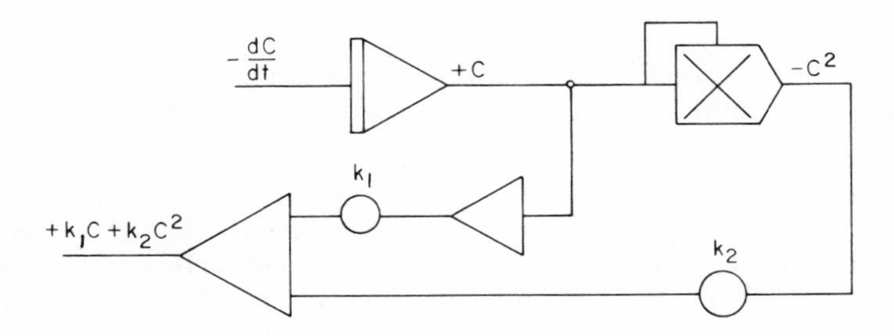

(4) We have now produced the terms equal to the highest derivative which was assumed at the outset; therefore, close the loop.

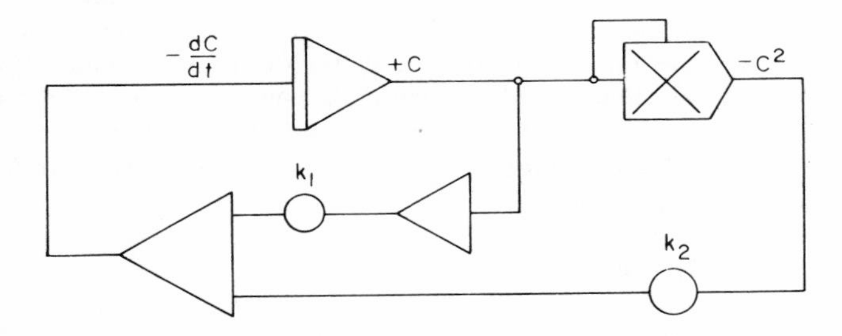

(5) Set the appropriate initial conditions. Only one is necessary here—the initial concentration

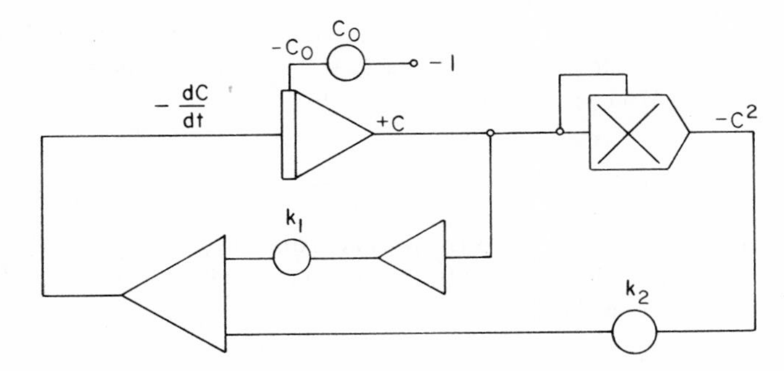

(6) Upon initiation of the program all of the variables in the equation will change according to the differential equation.

In the case of this particular example, and in general, it is useful to anticipate the sign of the variable to be displayed and arrange the sign of the derivatives to obtain it. Thus, +C would be more useful for display purposes than -C, since the inversion that attends analog integration -dC/dt was chosen as the derivative.

The simplicity of this approach and the immediate mastery of a forbidding differential equation is a delight and a marvel to the beginner. The feeling is well described in the words of the first person to experience it, Lord Kelvin:

"So far had I gone and was satisfied, feeling I had done what I wished to do for many years. But then came a pleasing surprise. Complete agreement (existed) between the function fed into the machine and that given out by it .... Thus I was led to a conclusion which was quite unexpected; and it seems to me very remarkable that the general differential equation of the second order with variable coefficients may be rigorously, continuously, and in a single process solved by a machine." [12]

Time is the only independent variable that an analog computer can use. What, then, if the differential equation is not time-dependent? One simply makes a scaling equivalency between time and whatever the problem's independent variable is. For example, consider the time-independent or stationary state solutions of the Schrödinger equation. Here the independent variable is distance. Then one simply makes the conversion from time to distance. Specifically consider the quantum harmonic oscillator form of the Schrödinger equation [13, 15]:

$$\frac{h^2}{8\pi^2\mu}\frac{d^2\psi}{dx^2} + (E - \tfrac{1}{2}kx^2)\psi = 0. \tag{9}$$

Then, rearranging and substituting constant values we have

$$\frac{d^2\psi}{dx^2} = 1.49 \times 10^{-2}\,\mu kx^2\psi - 5.95 \times 10^{-2}\,\mu E\psi, \tag{10}$$

where x is in angstroms, μ is the molar reduced mass, k is in dyne $\cdot$ cm^{-1}, and E is in cm^{-1}. Choosing HCl as a specific molecule ($k = 4.8 \times 10^5$ dyne cm^{-1} and $\mu = 0.9790$ g) we find that Eq. (10) becomes

$$\frac{d^2\psi}{dx^2} = 7.00 \times 10^3 x^2\psi - 5.83 \times 10^{-2} E\psi. \tag{11}$$

Now we are ready to make the transition from the problem's independent variable, x (Å), to the machine's independent variable, 6 (sec). The linear relationship between the two is simply

$$t(\text{sec}) = ax\ (\text{Å}), \tag{12}$$

where a is the scale factor. Making the transformation

$$\frac{d^2\psi}{dt^2}\left(\frac{dt}{dx}\right)^2 = 7.00 \times 10^{-3}\frac{t^2}{a^2}\psi - 5.83 \times 10^{-2} E\psi, \tag{13}$$

we obtain Eq. (14):

$$\frac{d^2\psi}{dt^2} = 7.00 \times 10^{-3}\ t^2\psi(a^{-4}) - 5.83 \times 10^{-2} E\psi(a^{-2}). \tag{14}$$

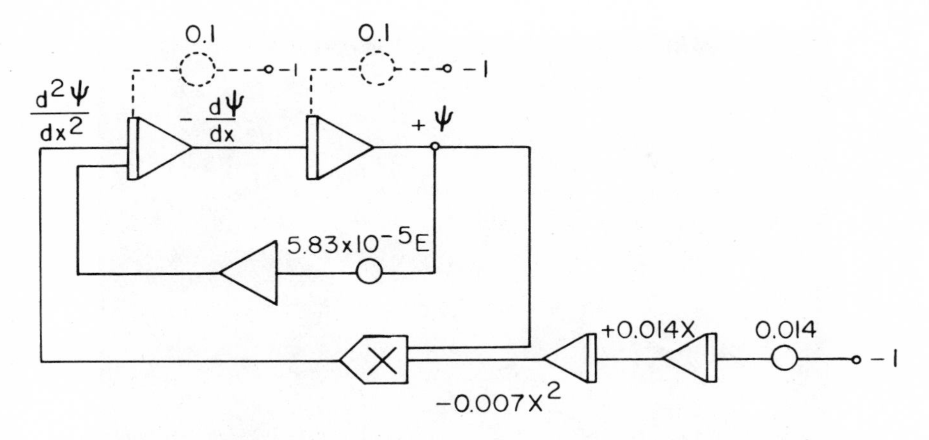

Fig. 1. Scaled program for HCl as a quantum harmonic oscillator.

The coefficients for $t^2\psi$ and $E\psi$ should be less than 1 so that the solution does not proceed too rapidly. So try $a = 10\sqrt{10}$ sec Å^{-1}. Then

$$\frac{d^2\psi}{dt^2} = 7.00 \times 10^{-3} t^2\psi - 5.83 \times 10^{-5} E\psi. \tag{15}$$

The program for the quantum vibrations of HCl is shown in Fig. 1.

If the program is run on the x1 time scale (RC for integrator is 1 sec) then 1 second represents 1/31.6 ($1/10\sqrt{10}$) of an angstrom.

The speed of the analog system which was described earlier is demonstrated in Fig. 2 where computer solutions of Eq. (9) are displayed. Each of these frames required 1 msec to trace. It is important to note that because of the parallel operation of the analog computer all solutions are simultaneously available. By rapid multiplexing, then, all of the plots shown in each frame were plotted in 1 msec.

2. Division by Zero

It is an important and useful concept in analog computation to recognize that any nth-order differential equation can be expressed as n first-order differential equations. Take, for example, the Legendre differential equation (4). This can be expressed as

$$\frac{dy}{dx} = y', \tag{16}$$

$$(1 - x^2)\frac{dy'}{dx} - 2xy' + n(n + 1)y = 0, \tag{17}$$

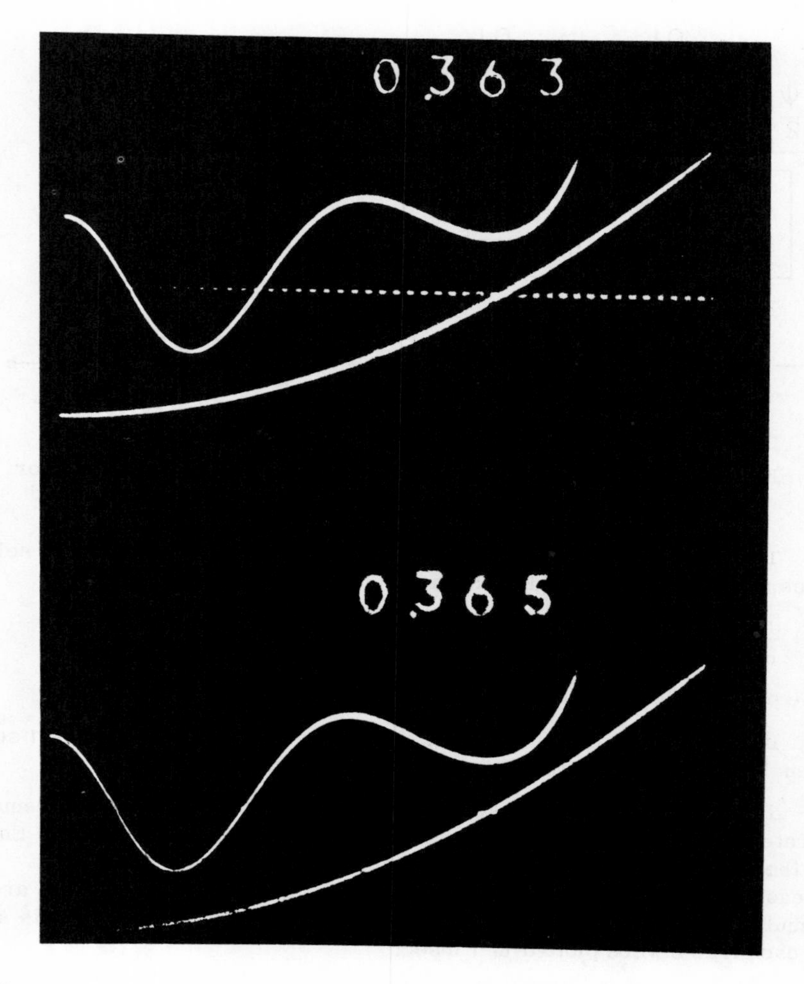

Fig. 2. Solutions of quantum harmonic oscillator (from film loop).

This concept is a useful one to remember for three important reasons. It is the basis for phase plane plots (an important tool for analyzing a differential equation which will be described later) and it forces the programmer to make a decision about each initial condition in the problem. It is also useful in surmounting a problem exemplified by the Legendre equation. If we apply the general method to the Legendre equation we have a problem at the end of the first step:

$$\frac{d^2y}{dx^2} = \frac{2x}{(1 - x^2)}\frac{dy}{dx} - \frac{n(n + 1)}{(1 - x^2)}y. \tag{18}$$

In this form at x = 1 the equation has a singularity, i.e., it becomes infinite. The problem comes in dividing by $1 - x^2$. The original equation did not suffer from this. What we would like to do is perform the solution by the general method and at the same time avoid division. There is a way to do this — it is based on the following considerations [4a, 16]. Any first-order differential equation can be considered as a quotient:

$$\frac{dy}{dx} = \frac{F(x, y)}{G(x, y)}. \tag{19}$$

The following transformation can then take place:

$$\frac{dy}{dt} = KF(x, y), \qquad \frac{dx}{dt} = KG(x, y), \tag{20}$$

where K is a constant. Note that the solution of dy/dt or dx/dt is straightforward with analog integrators, and the transformation solves the division that produced the singularity. Let us examine the solution in detail. First, it is necessary to transform the original differential equation into first-order equations. That has been done in (16) and (17). Then, generating the time derivatives and using unity coefficients, we have

$$\frac{dy'}{dt} = 2xy' - n(n + 1)y \tag{21}$$

$$\frac{dy}{dx} = y'; \qquad \frac{dx}{dt} = 1 - x^2 \tag{22}$$

and by the chain rule

$$\frac{dy}{dt} = \frac{dy}{dx}\frac{dx}{dt} = (1 - x^2)y'. \tag{23}$$

We can solve for y and x in the following manner. First, we solve the three time derivatives to produce y, y′, and x then we combine to produce the terms these derivatives equal, as defined by (21), (22), and (23). Using the initial conditions [4a],

n =	0	1	2	3	4	5	6
y(0) =	1	0	-1/2	0	3/8	0	-5/16
$\left(\frac{dy}{dx}\right)$ =	0	1	0	-3/2	0	15/8	0

and the program in Fig. 3, the first seven Legendre polynomials can be generated. They are shown in Fig. 4.

The "price" paid for the transformation is that the solution is not plotted linearly as a function of time. In the regular solution the x axis

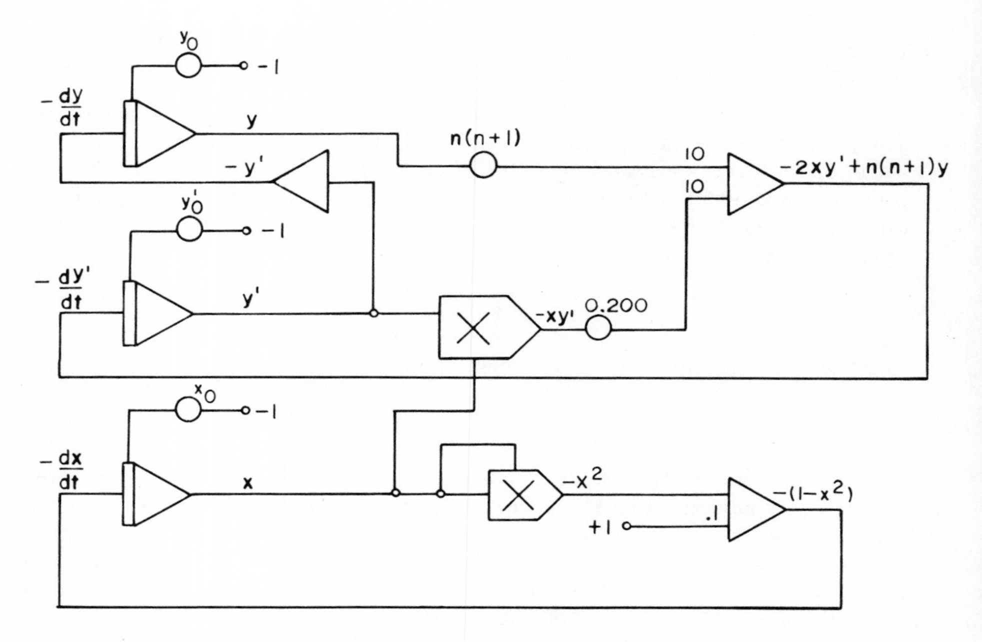

Fig. 3. Program for solution of the Legendre differential equation using division-avoiding technique.

would have been generated by integrating a constant so that y would be plotted as x moved at a constant rate. The division-avoiding transformation still produces the same graph, but, obviously, dx/dt is not a constant. This presents no problem in the final plot, which may merely require some extra attention in production. For example, in the transformation for Poisson's equation

$$\frac{1}{r^2}\frac{d}{dr}\left(r^2\frac{dy}{dr}\right) = \frac{-4\pi r}{\epsilon} = -K \tag{24}$$

the transformed equations are

$$\frac{dy}{dr} = y', \tag{25}$$

$$\frac{dy'}{dt} = \frac{-Kr - 2y'}{r}, \tag{26}$$

$$\frac{dr}{dt} = r. \tag{27}$$

Note that the r axis will not move initially unless a slight initial condition is set on the dr/dt integrator to get it started. Clearly, once integration begins, it will pick up speed.

Fig. 4. Legendre polynomial solutions.

3. Canonical Method [11b]

Even with division-avoiding transformations there are some differential equations that the general method cannot solve. For example, consider a general ordinary differential equation

$$\frac{d^n y}{dt^n} + a_{n-1} \frac{d^{n-1} y}{dt^{n-1}} + \cdots + a_o y = b_m \frac{d^m x}{dt^m} + b_{m-1} \frac{d^{m-1} x}{dx^{m-1}} + \cdots b_o x, \quad (28)$$

where x is the input function and y is the output function. The choice of input/output is quite arbitrary, as will be seen. The problem is that by the general method the $d^i x/dt^i$ terms can be integrated to solve for y, but there is no way to generate the derivatives of y. (Differentiation is to be avoided since it is very sensitive to noise.) The canonical and dummy variable methods are variations on the general method which provide a solution to such an equation. However, before these methods can be introduced we must investigate the concept of differential operators.

It is convenient when dealing with differo-integral equations to define an operator p with the following characteristics:

$$p = \frac{d(\,)}{dt}, \qquad p^n = \frac{d^n(\,)}{dt^n}, \tag{29}$$

$$p^{-1} = \int (\,)\, dt \tag{30}$$

With this notation then the general equation (27) becomes

$$p^n y + a_{n-1} p^{n-1} y + \cdots + a_o y = b_m p^m x + b_{m-1} p^{m-1} x \cdots + b_o x. \tag{31}$$

With this background we can develop the canonical method. Suppose we have an equation which is second order in y and first order in x:

$$p^2 y + a_1 py + a_o y + b_1 px + b_o x, \tag{32}$$

where a and b are constant coefficients. Then, rearranging so that the highest y derivative is by itself we have

$$p^2 y = b_1 px - a_1 py + b_o x - a_o y, \tag{33}$$

$$p^2 y = p(b_1 x - a_1 y) + b_o x - a_o y. \tag{34}$$

Integrating twice (multiplying through by p^{-2}) we have

$$y = p^{-1}(b_1 x - a_1 y) + p^{-2}(b_o x - a_o y). \tag{35}$$

(1) In summary then the first step in the canonical method is to arrange the differential equation so that the output variable stands by itself.

(2) Then assume that x will be available as an input to a series of integrators and that y will be the output. (Note the contrast with the general method, where the highest-order derivative was assumed):

(3) Then solve for the terms in y. For the first term we have

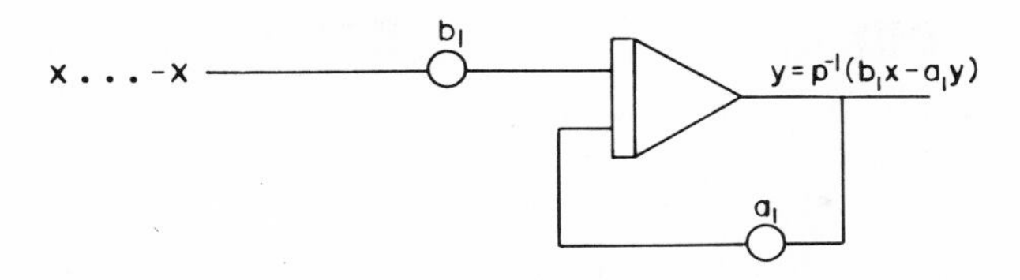

and for the second term we have

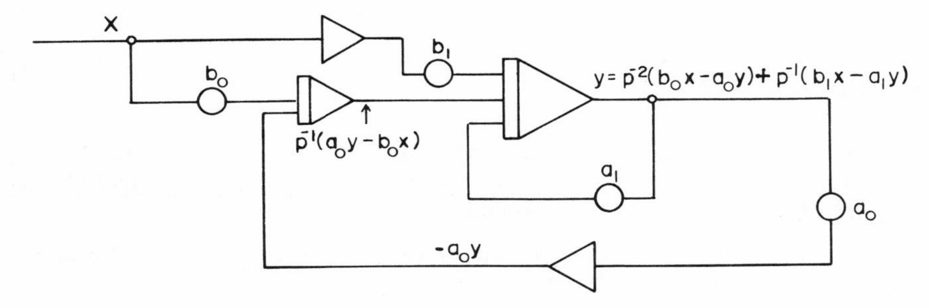

(4) Assume x, the forcing function, can be obtained. This program will then demonstrate the behavior of y as formalized by Eq. (31) for any forcing function or input x.

The systems described by equations are typically lead-lag systems where y follows x with a phase and amplitude difference between them, for example, a damped oscillator being driven by a forcing function:

$$p^2x + apx + bx = p^2y + cy. \tag{36}$$

It is important to note that the coefficients can be variables in the canonical method. This is not the case for the dummy variable (DV) method, which is easier, although it is limited to constant coefficients.

4. Dummy Variable Method [11c]

This method starts with an equation containing a dummy variable, solvable by the general method, i.e., containing no derivatives of x, and then modifies this dummy variable as a bridge is made to the desired equation.

Thus, to solve Eq. (31) the dummy variable basic equation would be

$$p^2y_d + a_1py_d + a_o y_d = x, \tag{37}$$

where y_d is the dummy variable. To build the zeroth derivative x term in Eq. (31) multiply (37) by b_o to give:

$$b_o p^2 y_d + b_o a_1 p y_d + b_o a_o y_d = b_o x. \tag{38}$$

For the first-order derivative in (31) multiply (37) by $b_1 p$ to give:

$$b_1 p^3 y_d + b_1 a_1 p^2 y_d + b_1 a_o p y_d = b_1 p x. \tag{39}$$

Then add (38) and (39) together and obtain:

$$p^3 b_1 y + p^2(b_o + b_1 a_1) y_d + p(b_o a_1 + b_1 a_o) y_d + b_o a_o y_d = b_o x + b_1 p x. \tag{40}$$

Now in order for the dummy variable version (40) of (31) to equal (31) it is necessary that

$$\frac{y}{b_o + b_1 p} = y_d. \tag{41}$$

(1) The first step in the dummy variable method, therefore, is to obtain a relationship between a dummy variable (based on an equation soluble by the general method) and the actual variable desired.

(2) Set up the solution of the dummy variable equation. Thus, for the solution of Eq. (37) we have the following program:

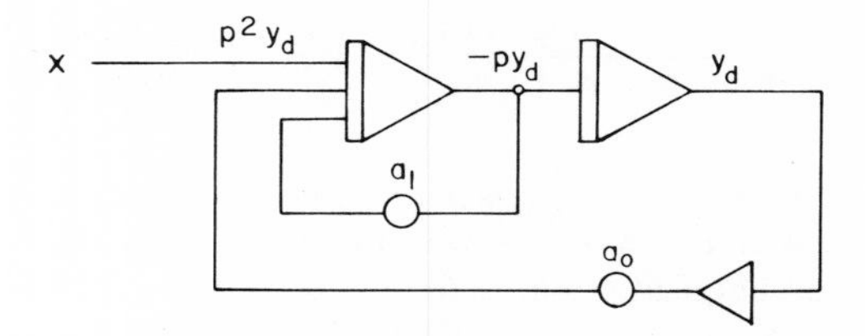

(3) Then, using the expression which relates y to yd, we generate y:

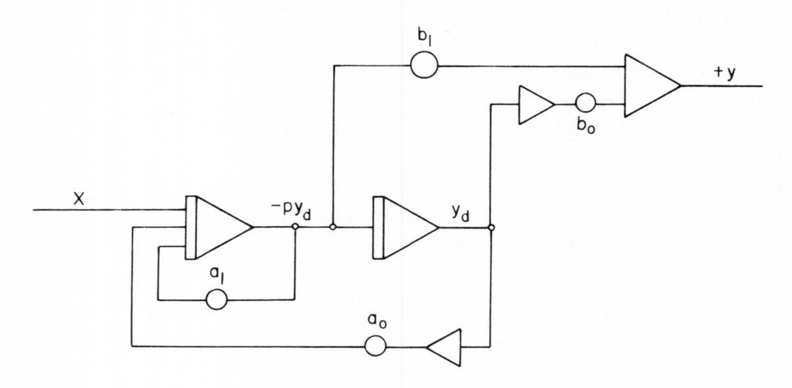

5. Phase Plane Plots

The choice of coordinates for displaying the solution of a differential equation can be very important. This is particularly true when dealing with coupled differential equations. (Since an nth-order differential equation can be considered as n coupled first-order equations, any second- or higher-order differential equation is, in effect, a coupled system). Consider the coupled pair

$$\dot{x} = k_1 x - k_2 xy, \tag{42}$$

$$\dot{y} = -k_3 y + k_4 xy, \tag{43}$$

proposed by Lotka [17] to explain ecological oscillations and, more recently, oscillating chemical reactions [18]. x/y can be considered prey/predator or host/parasite in the general kinetic scheme

$$x \xrightarrow{k_1} 2x,$$

$$x + y \xrightarrow{k_4} 2y,$$

$$y \xrightarrow{k_3} .$$

Depending on the magnitude of $x(0)$, $y(0)$, k_1, k_2, k_3, k_4 the behavior of x and y can be either unstable, stable and oscillating, or stable and fixed. An $x(t)$ or $y(t)$ plot versus time is not nearly as revealing as a plot of $x(t)$ vs $y(t)$. Figure 5 demonstrates this. This plot is called a phase plane plot. It is very informative, for at a glance it describes the behavior of the equations for a variety of parameters.

B. Partial Differential Equations

Analog and hybrid computers are quite effective in dealing with partial differential equations. There are a variety of techniques and a great deal has been written about them [19-25]. This section will introduce only a selected few. For more information the reader should consult the References. Solution of a partial differential equation requires that we perform integrations with respect to more than one variable. Since the analog computer has only one independent variable, namely, time, it is clear that continuous solutions are possible only for one variable. Algebraic interpolation over discrete intervals then becomes necessary.

Another important feature of many partial differential equations is that they are boundary value problems, that is, it is necessary that, at some time t after the computation begins, the functions have a specific value. This is to be contrasted with the initial condition type of equation treated so far in which only the initial value need be specified. The boundary value solutions require a reiteration, e.g., changing the initial condition until the boundary value is reached.

Fig. 5. Phase plane plot of Lotka prey predator system.

1. Boundary Value

As an example consider the linear diffusion equation

$$\frac{\partial C}{\partial t} = D\frac{\partial^2 C}{\partial x^2}, \tag{44}$$

which for concentration is known as Fick's Second Law and plays an important role in the theory of polarography [26]. Replacing concentration with temperature we arrive at an expression describing thermal conductivity. This can be stated as either a boundary value problem or an initial condition problem. First, as a boundary value problem, consider the following situation where flow takes place across the permeable membrane ($x = 0$), but no flow occurs across the barrier at $x = L$.

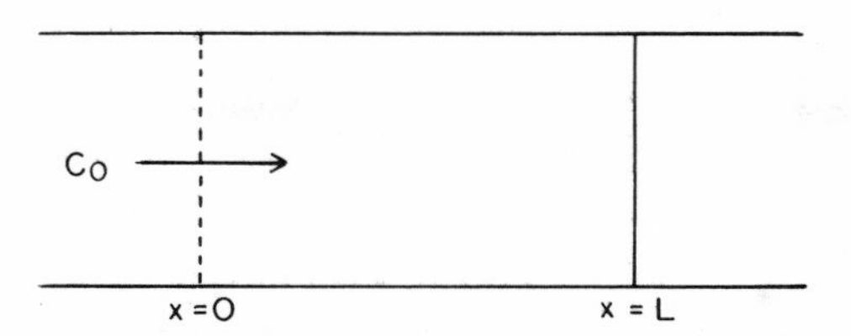

For a boundary condition problem we are only interested in the concentration profile that satisfies the conditions and no others. In this case a separation of variables is possible to give a single solution of interest [27-29]. Suppose the boundary conditions for C(x,t) are

$$C(o,t) = C_o \quad \text{and} \quad \left(\frac{\partial C}{\partial x}\right)_{L,t} = 0.$$

We can simplify the solution by first transforming to dimensionless parameters $X - x/L$, $\tau = Dt/L^2$ and then assuming that $C(X,\tau)$ is separable into independent space and time variables:

$$C(X,t) = f(X)g(t). \tag{45}$$

This permits a separation of X and τ for all values

$$f\left(\frac{dg}{d\tau}\right) = g\left(\frac{d^2 f}{dX^2}\right), \tag{46}$$

$$\frac{1}{g}\left(\frac{dg}{d\tau}\right) = \frac{1}{f}\left(\frac{d^2 f}{dX^2}\right), \tag{47}$$

only if the left- and right-hand sides of (46) are equal to a constant, call it K. Then

$$\frac{dg}{d\tau} = -Kg, \tag{48}$$

$$\frac{d^2 f}{dX^2} = -Kf. \tag{49}$$

Equation (49) is an eigenvalue equation; the eigenvalue -K is adjusted until the boundary conditions are met. Equation (49) is also a differential definition of a sine/cosine function with frequency $\sqrt{K}$.

2. Initial Condition

In the polarography case we have a depletion at the electrode which could be considered as a permeable membrane with C = 0 on one side and with a concentration C_o throughout the tube initially and $C_o = C$ at the electrode surface.

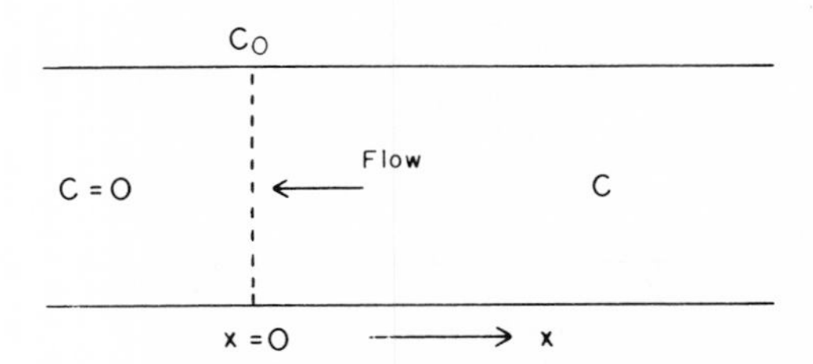

Then, as electrolysis proceeds, the concentration gradient extends further and further back into the tube. This then is an initial condition partial differential equation.

The discrete space-continuous time solution of the initial condition version of Fick's equation (44) will be demonstrated. By discrete space we mean that the algebraic approximation

$$\frac{\partial C_{1/2}}{\partial x} \cong \frac{C_1 - C_0}{x_1 - x_0} \quad \text{(first derivative for interval between 0 and 1)}$$

will be used for slices along the tube. Suppose then that we divide the tube into four slices (more for better definition) along the x-axis. The derivatives would then fall in the following manner:

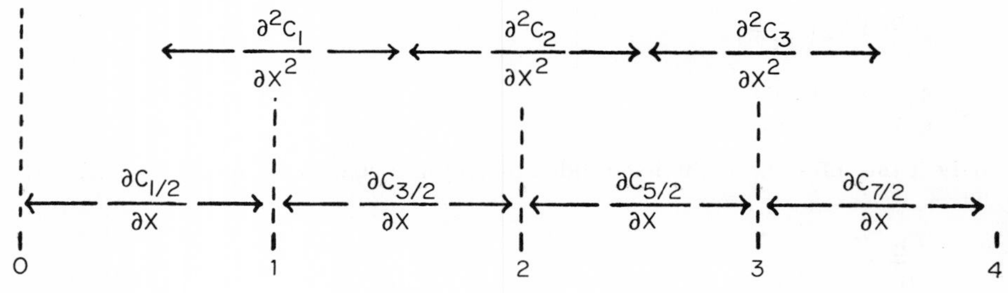

and

$$\frac{\partial^2 C_1}{dx^2} \cong \frac{(C_2 - C_1)/\Delta x - (C_1 - C_0)/\Delta x}{x_{3/2} - x_{1/2}}$$

$$= \frac{\partial C_{3/2}/\partial x - \partial C_{1/2}/\partial x}{\Delta x}.$$

(Δx assumes equal spacing.)

Figure 6 shows the program for implementing the equation. This program is rather consumptive, requiring an integrator, two amplifiers, and four coefficient potentiometers per final point. A great economy can be achieved by considering the derivative from a central difference point of view. Consider some point x along the tube with points placed at $\pm\Delta x$ to either side of it.

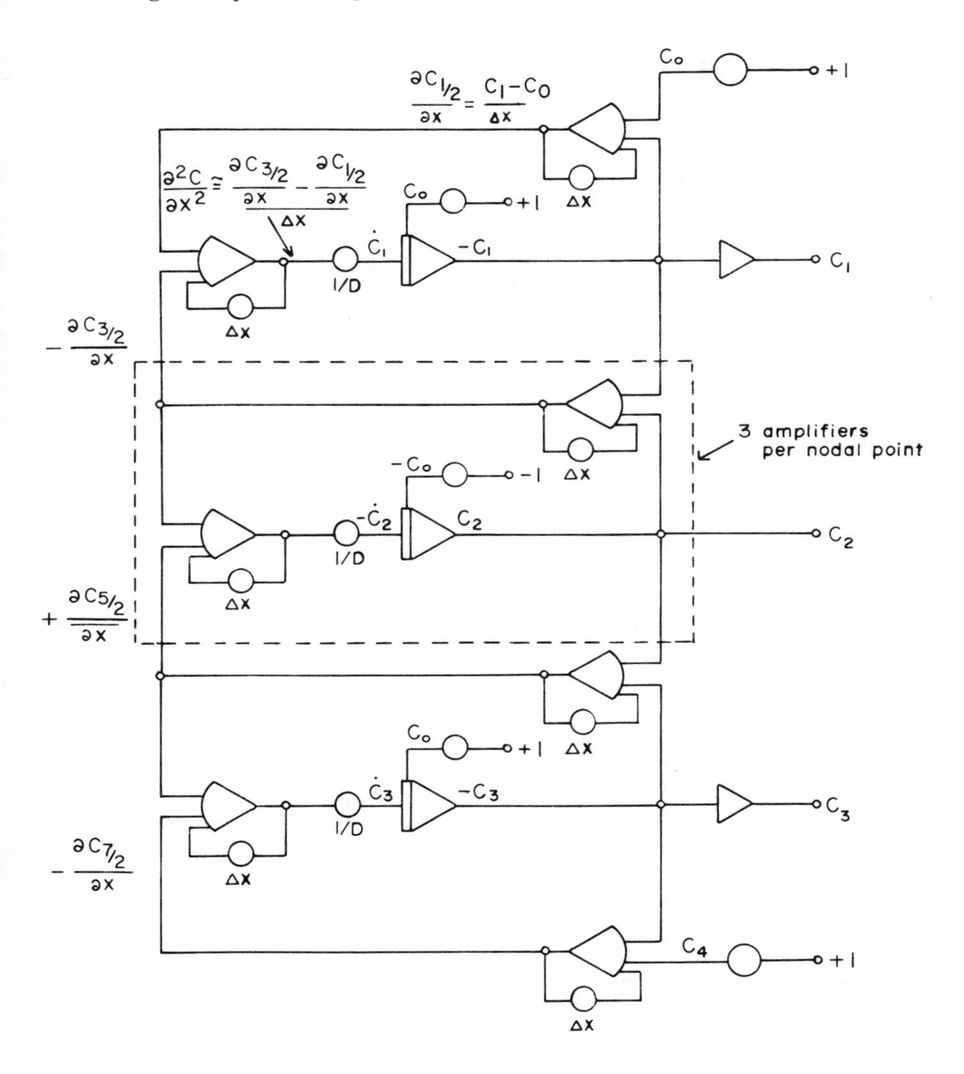

Fig. 6. Program for four-point solution of Fick's second law.

x − Δx　　　x　　　x + Δx

$$\frac{C_{x+\Delta x} - C_x}{\Delta x} = \frac{\partial C_{x-\frac{1}{2}\Delta x}}{\partial x} \qquad \frac{\partial C_{x+\frac{1}{2}\Delta x}}{\partial x} = \frac{C_x - C_{x-\Delta x}}{\Delta x}$$

Then the second central difference is given by

$$\frac{\partial^2 C}{\partial x^2} \cong \frac{\partial C_{x+\frac{1}{2}\Delta x} - \partial C_{x-\frac{1}{2}\Delta x}}{\Delta x} \cong \frac{C_{x+\frac{1}{2}\Delta x} + C_{x-\frac{1}{2}\Delta x} - 2C_x}{\Delta x^2}. \tag{50}$$

Now assign an integrator to each point and assume that the output is the concentration for that point (with alternating sign), as shown in Fig. 7.

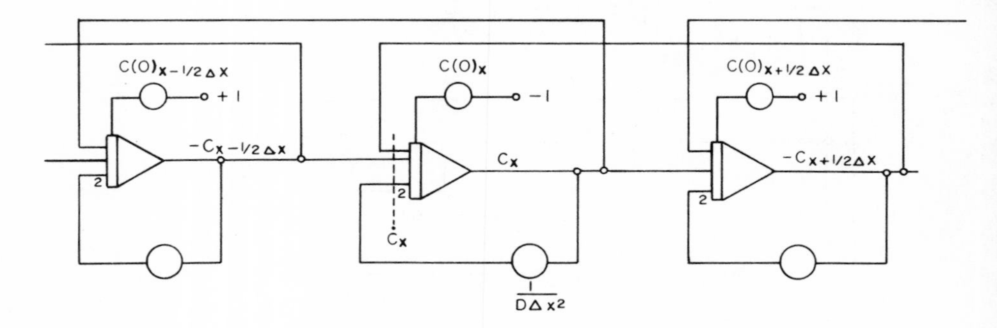

Fig. 7. Efficient program for solution of Fick's second law.

If we add a potentiometer to each output to define $1/D\Delta x^2$, then the outputs of the integrators on either side, when summed with $2C_x$, the output of the central integrator, are equal to $|\partial C/\partial t|$. This method is much easier to set up and requires less than half the analog components. It has the disadvantage that every other point has a negative value, but an inverter for display purposes easily solves that. The solution of a ten-point simulation (for $D = \Delta x = 1$) is shown in Fig. 8. All plots were produced simultaneously (parallel operation) in 3 msec.

Other classical techniques for solving partial differential equations using discrete time-discrete space and Monte Carlo methods are described elsewhere [19]. However, there is a new method made possibly by the hybrid techniques and new hybrid equipment [22]. The new device is simply the digital coefficient potentiometer (also called digital coefficient unit, DCU, or digital to analog converter, DAC). With this device we can have any number of nodal points using only one integrator. It operates as shown below:

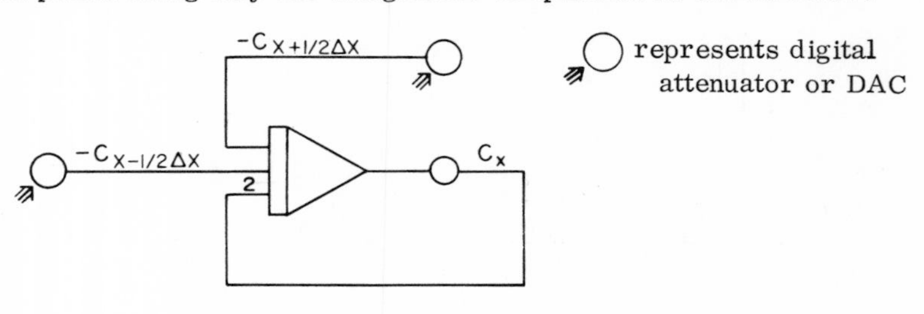

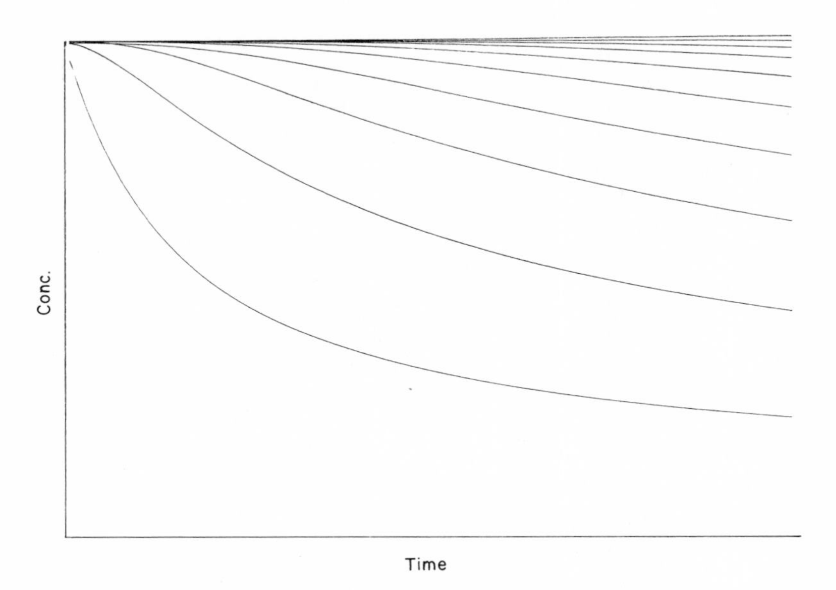

Fig. 8. Plot of ten-point solution of Fick's second law.

Let us begin by assuming that $C_{x-\frac{1}{2}\Delta x}$ is the first nodal point (C_0) and that $C_{x+\frac{1}{2}\Delta}$ is C_2, then C_1 is C_x. Set C_2 to C, the concentration in the bulk of the tube, then set $C_0 = 0$, integrate for a short interval, then put the integrator into HOLD, freezing the integration. Plot the output of the integrator as the value of C_1 at a time $\Delta\tau$ after diffusion has begun. Perform an A/D conversion on C_1 and then set DCU $C_{x-\frac{1}{2}\Delta x}$ to this value. Then go into OPERATE for another $\Delta\tau$. During this interval we have effectively moved the integrator to the next node along the x axis so that when the computer next goes to HOLD the output is C_2 corresponding to the same time for which C_1 was just calculated. This process continues for as many concentration nodal points as are necessary to reach the boundary at the end of the tube. At this point we have a C vs x profile which could have several hundred steps for a time corresponding to $\Delta\tau$ from $t = 0$. However, since the C_2 that we used for computing C_1 was an approximation, the profile may be in error, so we repeat the whole process, but this time use for C_2 not C but the value for C_2 we calculated the first time. We keep repeating this until further iterations produce no change in profile. For a hundred interval profile about 50 iterations are required, as shown in Fig. 9.

To get the next concentration gradient profile (C vs x) for $t = 2\ \Delta\tau$ (or any arbitrary time) the integrator would be reinitiated to $C_2 = C$ and then put

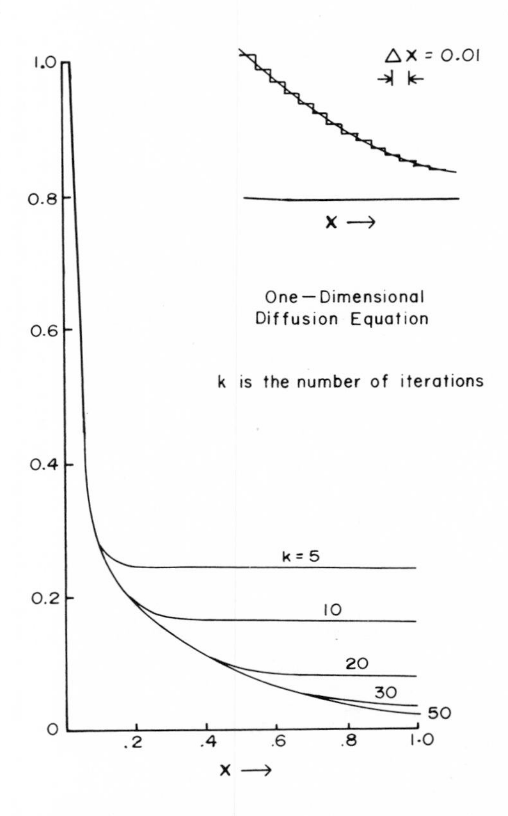

Fig. 9. Iterative concentration profiles for some time after t = 0.

into HOLD 2 $\Delta\tau$ (or whatever time was chosen) after C_0 was set to zero. Thereafter, the OPERATE intervals would have to be $\Delta\tau$, the same as the first sequence, so that the nodal points are congruent. The same reiterative process then continues.

The time required to solve differential equations in this manner is critically dependent on D/A conversion rates. The economy in hardware more than compensates for the cost in time, for the number of integrators required for detailed profiles is large; for two- and three-dimensional equations cost becomes prohibitive.

There is still another way that the latter method can be implemented without DCUs, utilizing the logic capability available on an analog/hybrid machine — to use an integrator as a memory device. The next section will describe the logic capability of the modern analog/hybrid computer.

III. LOGIC

A. Typical Computer Complement

All current analog/hybrid computers contain a variety of logic elements which enable the computer to make decisions and operate under programmable control in a stand-alone configuration. The logic capability is greatly enhanced with digital computer control, although some of the logic capability then becomes redundant since it can be performed as effectively with software.

This section, like the introduction to the analog section, will assume that the reader is familiar with the operation of logic elements [30-32]. A typical complement of logic elements to be found on a fully expanded, medium-sized analog/hybrid computer is given in Table 5.

TABLE 5

Fully Expanded Logic Compliment for Medium-Sized Analog/Hybrid Computer

	AD/5[a]	680[b]	SD10[c]
Flip Flops	32	24	12
Gates	60	36	12
BCD counters	6	3	3
Monostable timers	0	6	3
Differentiators	16	6	-
Comparators	16	24	20

[a]Applied Dynamics.

[b]Electronic Associates, Inc.

[c]Systron-Donner.

In addition there is patchable control capability for integrator modes (INITIAL CONDITION, OPERATE, HOLD) and time scales, electronic switch, and relay and comparator outputs. The timers and flip flops are synchronized with a 1 MHz clock. The logic units, therefore, operate in synchronization and in parallel. From these logic units can be easily constructed the more complex ring counters, modulo, and sequencing units which are familiar to the reader. The important point here, however, is that these more sophisticated combinations are not an end in themselves but serve vital and sometimes unexpected computational functions.

B. Applications of Logic in Programming

It would be impossible to catalog all of the uses of the logic section of the analog/hybrid computer. A selection of examples will be given to give some indication of the possibilities and at the same time demonstrate the typical features of the logic elements. Four uses involving solution display and computation will be described.

1. Display-Computer Interfacing

Display strategies will depend on display devices, and the requirements of these should be examined before proceeding. Three signals will be available for any display device: the coordinates X and Y, and intensity Z. Basically, two types of display devices are available: x-y plotter and cathode ray tube. A useful feature on the plotter is the provision for remote pen up/down control. This enables a binary Z logic signal to completely control what is plotted. Since the T^2L logic is not compatible with the solenoid that controls the pen, a logic-driven relay (an option to be found in the logic patchpanel) is necessary.

One must proceed with caution in the selection of a CRT which will have Z-axis requirements that an analog computer can provide. An oscilloscope that has no way of controlling the intensity externally will have limited value as a computer display device. Oscilloscopes requiring ac-coupled Z signals are usable but awkward. A binary Z axis is quite useful; it is similar to an x-y plotter, but simpler in that it can be driven directly by T^2L logic. An oscilloscope that utilizes an analog Z axis is probably the most powerful display adjunct because it can be used in a binary mode or with an analog Z signal so that information can be inserted in the intensity pattern. For phase plane plots it is necessary that both x and y axes accept external inputs.

The logic signals that the programmer has at his disposal are straightforward. The INITIAL CONDITION (IC), OPERATE, and HOLD signals or their complement are available. Normally, the Z axis can be driven simply by the OPERATE BUS (OPB) or its complement. For more sophisticated graphs a flip flop can be enabled by pulses available from within the computer synchronously with the start of the operate signal, and then ANDed with the OPB. This output to the Z axis will produce a dotted or labeled line. Suppose that one wanted to plot a function during the computation only when it is within a certain voltage range. Then a comparator could set the range and the logic output from the comparator ANDed with OPB to give the desired Z signal.

2. Multiple Curve Display

The very high speed capability of the analog computer makes it possible to consider multiple trace displays without flicker. For example, one can perform consecutive computations displaying a new variable each time. At a rate of 1.5 msec per computation (0.5 msec IC, 1.0 msec OPERATE), nearly 20 different plots can be displayed before flicker becomes noticeable. This could be done using a ring counter constructed from flip flops. The flip flops available at the patch panel are rather sophisticated. For example, in the AD/5 a flip flop may be in either the LOAD, RUN (clock on), or STOP (clock off and flip flop in state immediately before RUN terminated) modes. The flip flop may set to either condition from a key switch on the console panel. The console key switch may be overridden by a logic "1" patched into the LOAD (LD) patch hole. The LOAD state, "1" or "0", can be set by a console key switch with patch hole LOAD set (LDS) overriding. The flip flops have the truth table shown in Fig. 10 [33a]. They are arranged in pairs with a common ENABLE, LOAD, CARRY OUT, and CARRY OUT COMPLEMENT.

If a new plot is to be obtained on each new computation we need a pulse at the beginning of the OPERATE signal. This is obtained from a leading edge differentiator, a typical element to be found in the logic section. Putting it all together we have the multiplexer shown in Fig. 11.

Of course, this multiplexer does not take advantage of the parallel operation of the analog computer. If we multiplexed fast enough we could display all 20 in one computation! To do this we would use clock pulses from the computer. Since all logic elements are synchronized with a 1-MHz clock we cannot control logic devices at that speed. Generally, 10-μsec pulses are the fastest available.

If there are n traces to be shared then each trace will have only 1/n of the entire trace plotted. However, if the multiplexing is asynchronous from one computation to another, the gaps will be filled in on subsequent computations, giving the viewer the impression of a solid line. The implementation of this is shown in Fig. 12.

For most electromagnetic deflection oscilloscopes 10 μsec represents the maximum speed anyway, since a few μsec are required to settle onto a new point. Figure 13 shows the high speed multiplexed version of Fig. 8.

3. Coordinate Grid Reference [33b]

A coordinate grid reference is often very useful in a graphical display. Although some oscilloscopes have graticules for this purpose, more often than not the graticule is not quite suitable for the display of interest, and further, there is usually a parallax problem. A simple Modulo Counter controlling the coefficients to an integrator makes it possible to generate a grid reference easily and quickly. A binary coded Modulo-X Counter is a

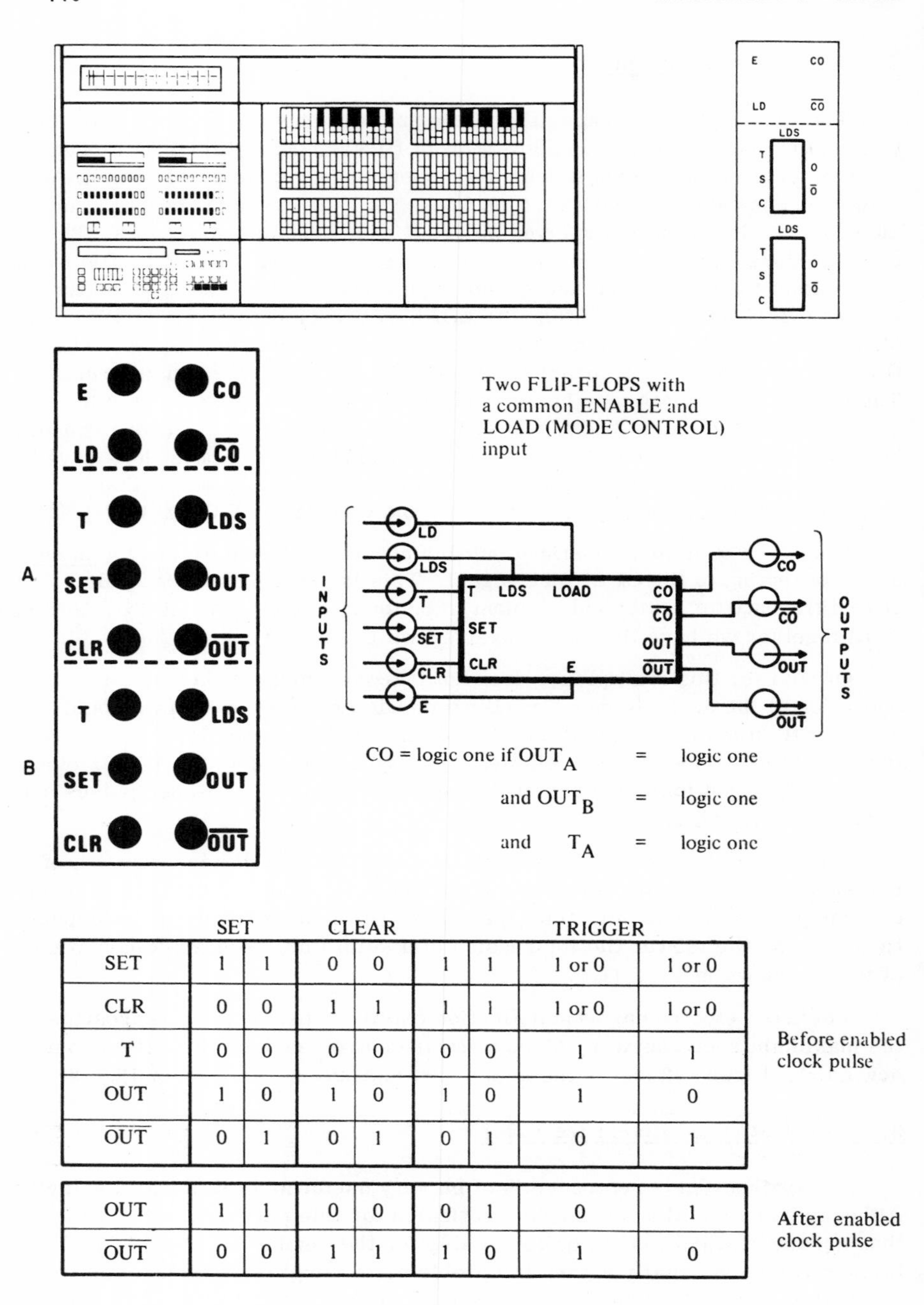

	SET		CLEAR				TRIGGER	
SET	1	1	0	0	1	1	1 or 0	1 or 0
CLR	0	0	1	1	1	1	1 or 0	1 or 0
T	0	0	0	0	0	0	1	1
OUT	1	0	1	0	1	0	1	0
$\overline{OUT}$	0	1	0	1	0	1	0	1

Before enabled clock pulse

OUT	1	1	0	0	0	1	0	1
$\overline{OUT}$	0	0	1	1	1	0	1	0

After enabled clock pulse

Fig. 10. Typical flip flop to be found on analog/hybrid computer.

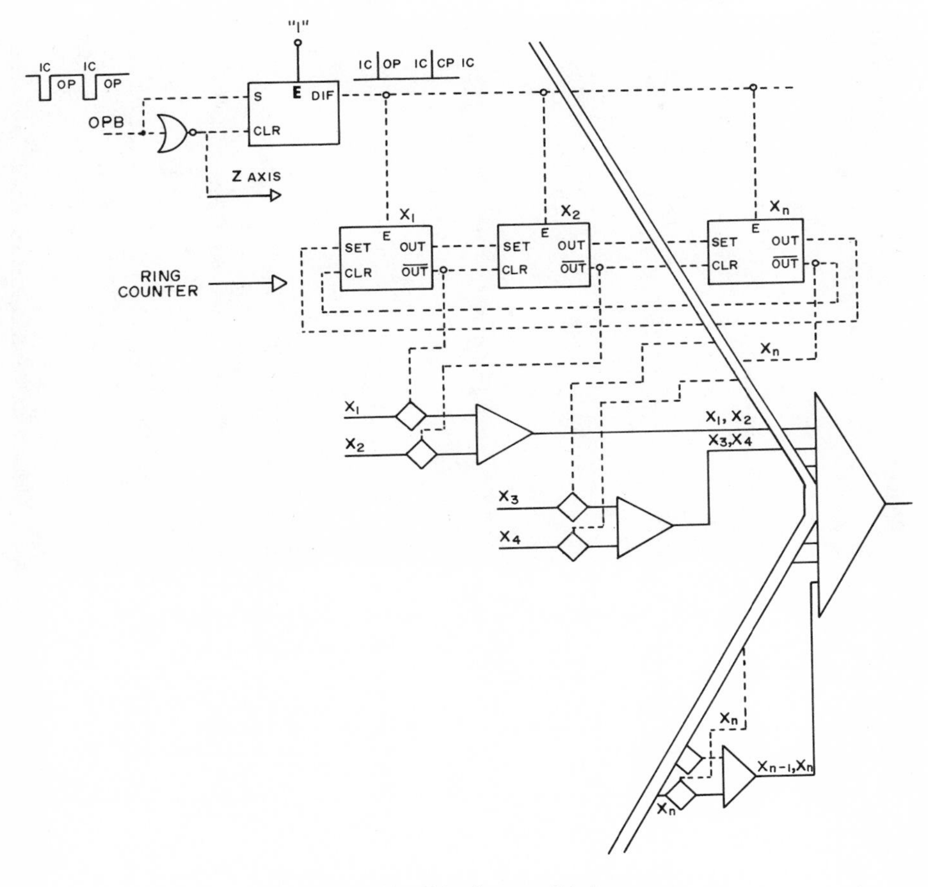

Fig. 11. Display multiplexer.

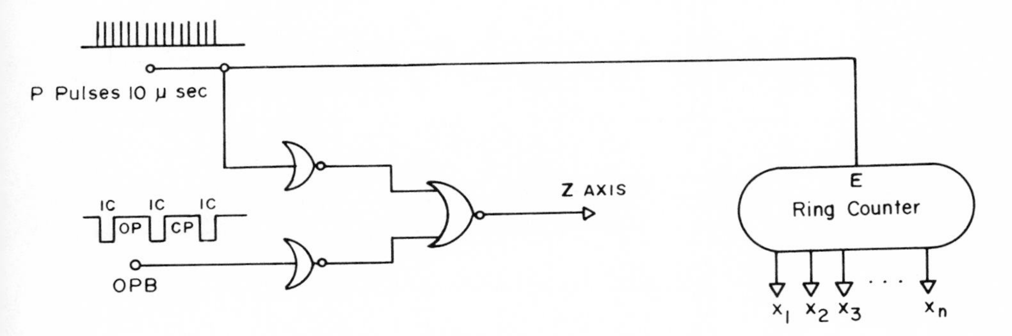

Fig. 12. High-speed asynchronous multiplexer.

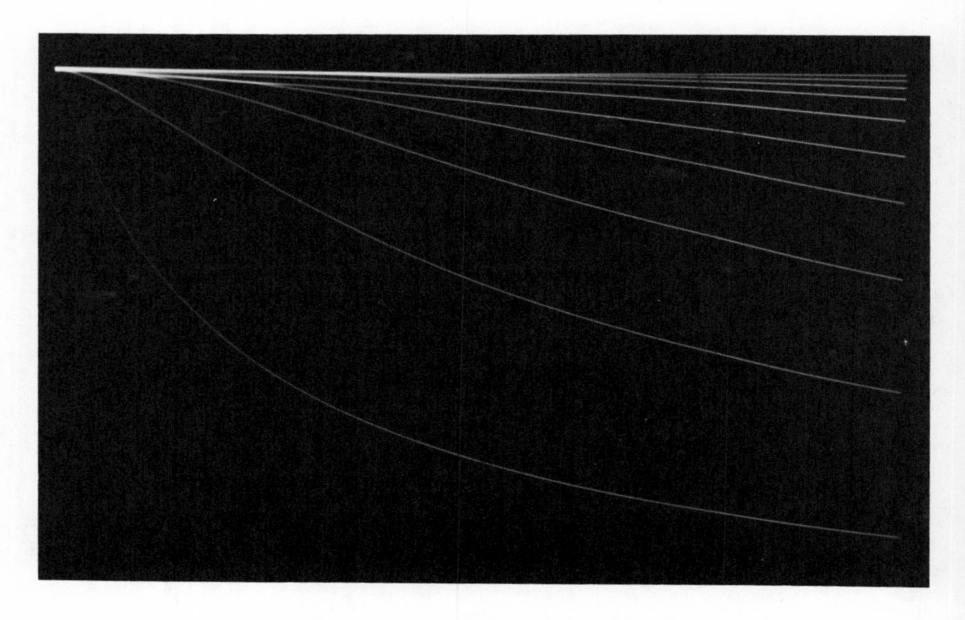

Fig. 13. High-speed multiplexed display of Fick's law.

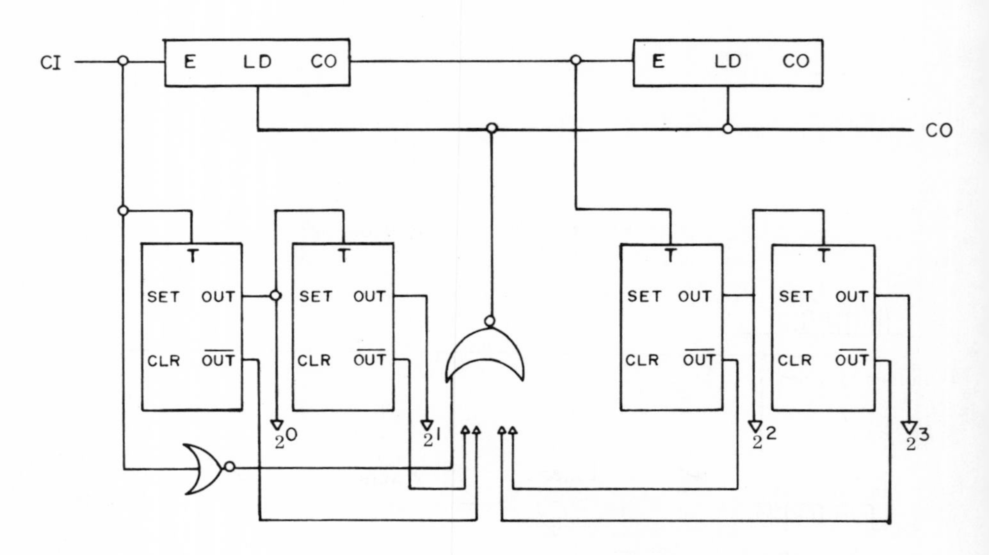

Fig. 14. Modulo-x counter.

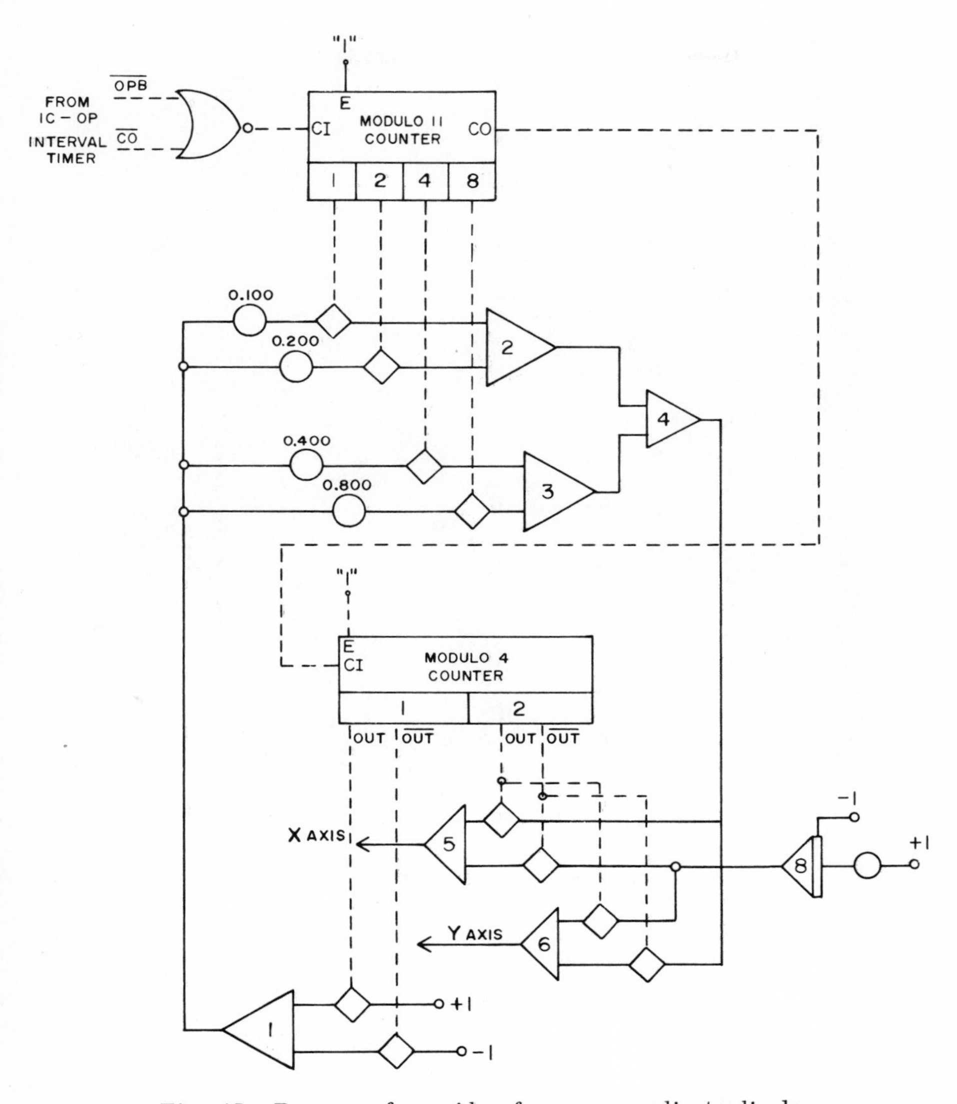

Fig. 15. Program for grid-reference coordinate display.

counter which will count to X in the familiar binary fashion. It is composed of flip flops as shown in Fig. 14 [33c].

If we then incorporate a Modulo 11 and Modulo 4 Counter as shown in Fig. 15 we can generate the grid shown in Fig. 16. The circuit works as follows. With each clock pulse that comes into the Modulo 11 Counter at CI

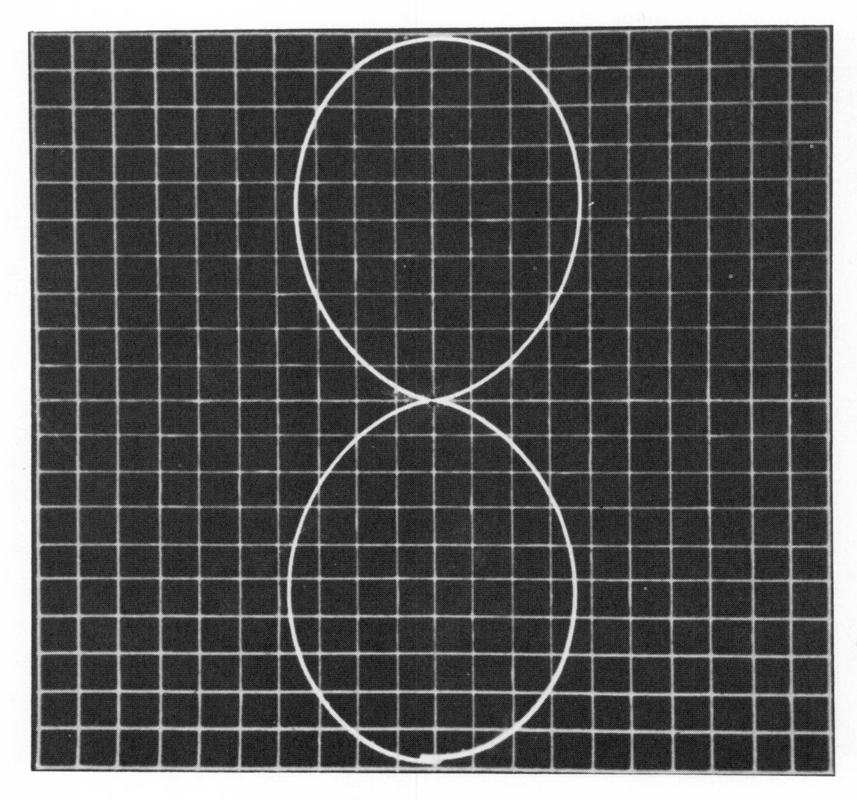

Fig. 16. Example of a grid reference pattern for ψ^2 of a particle on a ring.

(CARRY IN) the output of summer of counts 0, 0.1, 0.2, 0.3, 0.4, 0.5, 0.6, 0.7, 0.8, 0.9, 1.0, 0, -0.1, -0.2, ..., -1.0, 0, 0.1, 0.2, etc. Integrator 8 draws straight lines along the X axis from +1 to -1 first for the positive string of Y values, then for the negative string of Y values. At this point the Modulo 4 switches axes and repeats. The x = 0 = y axes are reinforced in intensity with double traces. The whole operation takes 44 traces, which at 100 μsec/trace represents 4.4 msec for the entire pattern. Consequently, this "graphic paper" unit can alternate with a problem and show no flicker. Since the displays are generated by the same analog equipment on the same surface, parallax is eliminated. Further, there are interesting possibilities for nonlinear graphs.

4. High Speed Sampler

During high speed repetitive operation it is sometimes useful to be able to plot on an x-y recorder the same graph that is being computed every

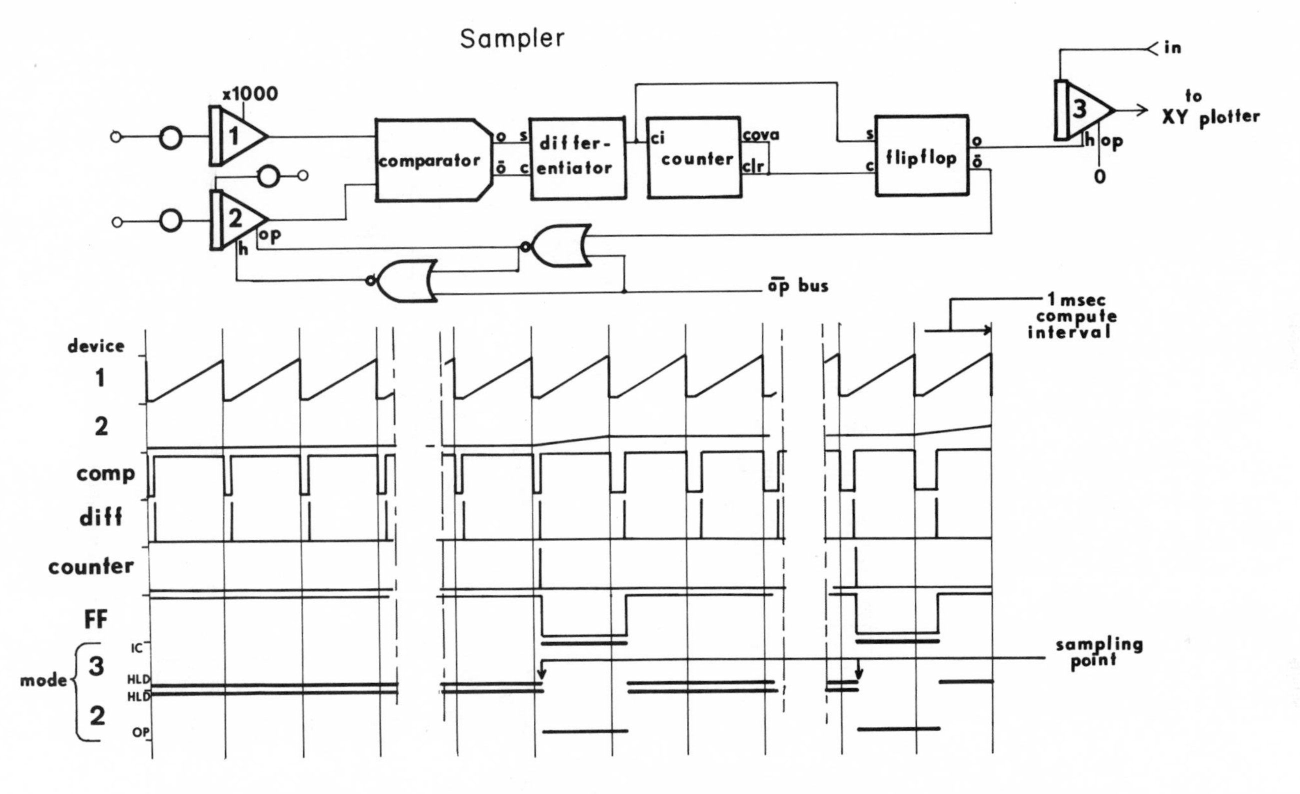

Fig. 17. Program for a high-speed sampler.

millisecond. This is particularly true for a time-shared system where one user would like a hard copy of the graph on the oscilloscope but at the same time does not want to slow down the central computer and make all other users wait. In other words, each user of the time-shared system is being serviced on a millisecond time-scale so that some users are viewing a flicker-free CRT display while others are recording their display on a recorder. What is needed is a narrow sampling aperture that slowly moves across the rapidly repeating function. The analog and logic circuitry for implementing this as well as the timing diagram are shown in Fig. 17.

Integrator 1 rapidly and linearly moves from 0 to +1 machine units during the course of each computation. Integrator 2 moves slowly from 0 to -1.

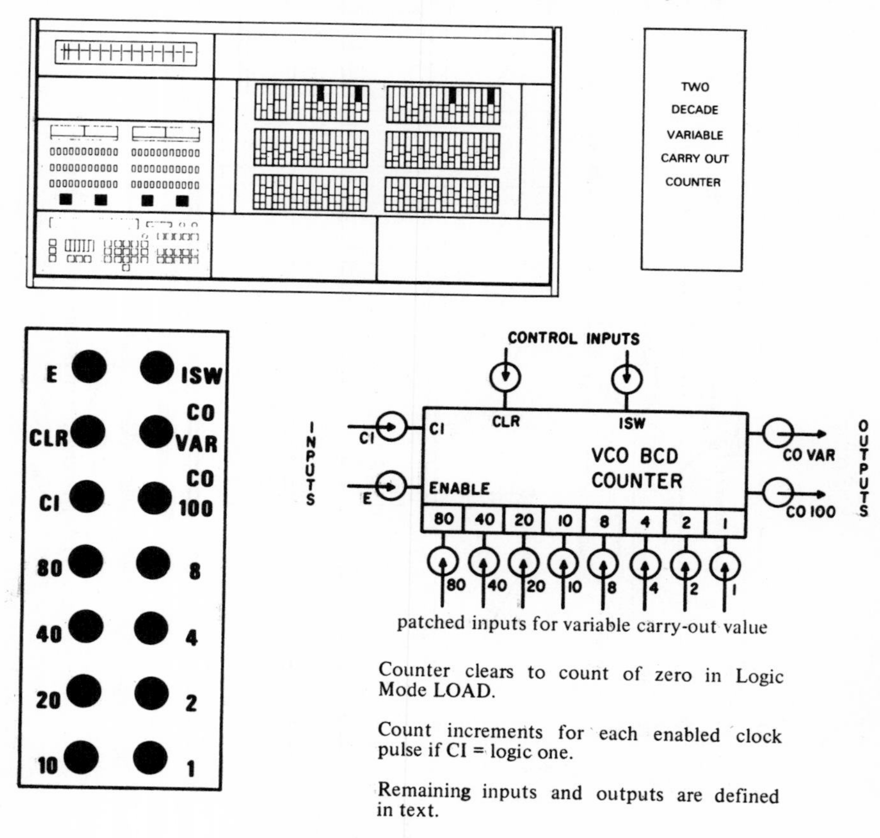

Patching a Variable Carry-Out BCD Counter

Fig. 18. Typical BCD counter to be found on analog/hybrid computer.

When the sum of the integrator outputs is ≥ 0, the comparator goes from low to high, which is differentiated. The differentiated signal sets a flip flop and increments a counter. The counter is a variable carry-out BCD counter. It has a variable count from 0 to 100 that can be set either from the control panel or through the patch panel holes shown in Fig. 18 [33d]. A logic "1" at ISW disables the panel control switches and activates the patched count. When the count reaches one less than the preset value a CARRY OUT (COVA) pulse is generated which, if connected to CLEAR (CLR), will clear the counter to zero. The flip flop is simultaneously cleared. The flip flop controls the mode of integrator 3, which is effectively a track-hold device. If we take the high-speed signal which we wish to follow and connect it to the IC patch hole of the integrator then in IC mode the integrator is tracking. When the output of the flip flop goes high the integrator goes into HOLD. The same flip flop also HOLDs integrator 2. Integrator 2 determines where the window is within the high-speed (1-msec) COMPUTE interval. The width of the window is less than 1 μsec. The counter determines how long integrator 3 holds a point so that the x-y plotter can plot it before moving on. In this manner it is possible to record in several seconds a graph which has been calculated in 1 msec several thousand times.

The last example of logic applications will demonstrate the three-dimensional capability of analog displays. Quite often it becomes necessary to deal with three-dimensional objects or graphical representations. Quite often a reaction coordinate or thermodynamic function can be clearly understood only if we have a three-dimensional surface to describe it [36-38]. There is a powerful method for representing three-dimensional contours that uses differential equations. This theory, coupled with some logic implementation, makes it possible to present contour surfaces and rotate them in space in real time.

5. Three-Dimensional Surfaces and Projections

Suppose that we have a transparent surface in three-space and we want to determine the shape of that surface. Furthermore, we have a rule that, whenever two surfaces intersect, the line that defines that intersection will be visible. Now, if we take a surface with which we are familiar and jab it into space a few times, the intersecting lines which appear, when taken together, will begin to describe both surfaces. If we use a plane as the exploratory surface and insert it in a parallel fashion we will produce the contours of the surface in question. Now let us consider how we do this mathematically.

Consider two surfaces $G(x, y, z)$ and $F(x, y, z)$ in three-space as shown in Fig. 19. In this figure the dotted line represents the intersection of the two surfaces. Take some point on that line and determine the gradient of G and the gradient of F. Then the cross product of these gradients is the tangential vector for that line. If we set the cross product equal to the time

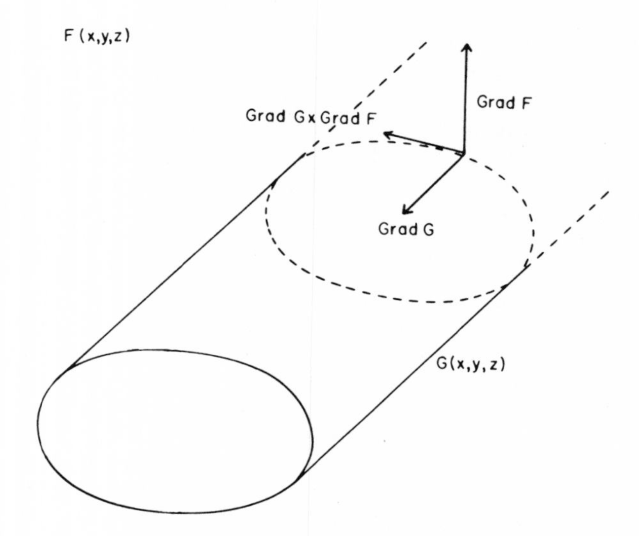

Fig. 19. Intersective surfaces.

derivations of x, y, and z then we can solve for the trajectory of this line through analog integration [34], i.e.,

$$\frac{d}{dt}\begin{pmatrix} x \\ y \\ z \end{pmatrix} = \begin{pmatrix} \eta \dfrac{\partial(F,G)}{\partial(y,z)} \\ \eta \dfrac{\partial(F,G)}{\partial(x,z)} \\ \eta \dfrac{\partial(F,G)}{\partial(x,y)} \end{pmatrix}, \tag{51}$$

where n is a nonzero constant and $\partial(G,H)/\partial(y,z)$ is the 2×2 Jacobian determinant.

Equation (51) gives the following correspondence between the time derivatives and the Jacobians:

$$\dot{x} = \begin{pmatrix} \dfrac{\partial F}{\partial y} & \dfrac{\partial F}{\partial z} \\ \dfrac{\partial G}{\partial y} & \dfrac{\partial G}{\partial z} \end{pmatrix}, \tag{52}$$

$$\dot{y} = \begin{pmatrix} \dfrac{\partial F}{\partial x} & \dfrac{\partial F}{\partial z} \\ \dfrac{\partial G}{\partial x} & \dfrac{\partial G}{\partial z} \end{pmatrix}, \tag{53}$$

$$\dot{z} = \begin{pmatrix} \frac{\partial F}{\partial x} & \frac{\partial F}{\partial y} \\ \frac{\partial G}{\partial x} & \frac{\partial G}{\partial y} \end{pmatrix}. \tag{54}$$

Take a specific example of a simple spherical surface to be sliced along the x axis:

$$F(x, y, z) = 0 = x^2 + y^2 + z^2 - a^2, \tag{55}$$

$$G(x, y, z) = 0 = x - b, \tag{56}$$

then we have

$$\dot{x} = \begin{vmatrix} 2y & 2z \\ 0 & 0 \end{vmatrix}, \quad \dot{y} = -\begin{vmatrix} 2x & 2z \\ 1 & 0 \end{vmatrix}, \quad \dot{z} = \begin{vmatrix} 2x & 2y \\ 1 & 0 \end{vmatrix}.$$

The analog solution is trivial (see Fig. 20).

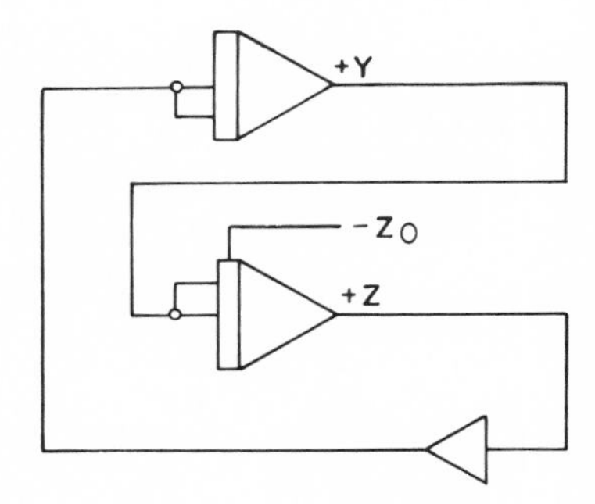

Fig. 20. Program for contour of sphere.

The initial conditions must be calculated algebraically from the equations for the surfaces, F(x, y, z) and ∂(x, y, z). Thus, in the case of the sphere-plane, the initial value for x (x_0) is given as b. Then if we choose y_0 to be zero, z_0 is determined. The program for doing this is shown in Fig. 21.

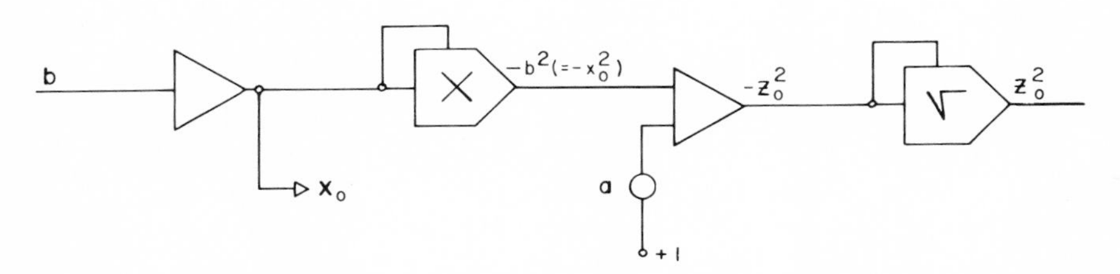

Fig. 21. Program for initial condition of spherical contours.

Contours can be obtained for other axes by choosing different planes, e.g. $y = b$ or $z = b$. A new slope can be explored by using a different $F(x, y, z)$.

For a given "b" value, x_o and z_o are determined ($y = 0$). If at the end of each computation we increment "b" to a new value, a new contour of the sphere is produced.

All that remains is the stepping of the "b" constant, done with a Modulo 8 which produces eight sections for one hemisphere along the positive x axis, then repeats along the -x axis. Figure 22 shows the program for this.

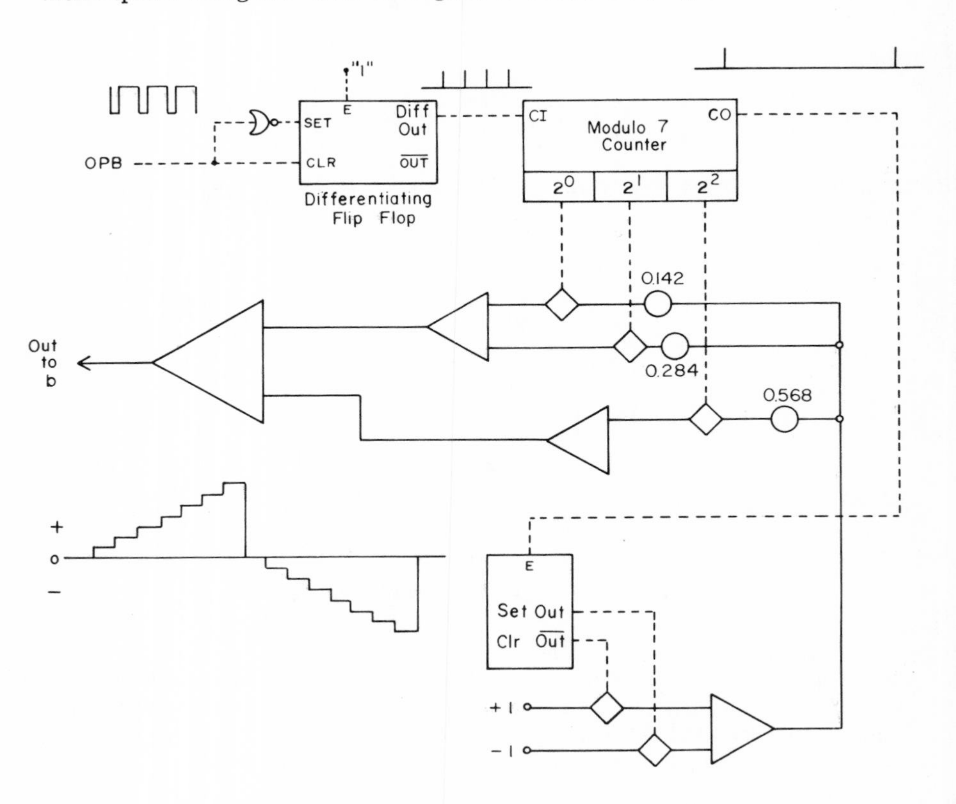

Fig. 22. Program for automatically sequencing equally spaced contours.

The following procedure enables us to take a point (x, y, z) and rotate it to a new point (x_1, y_1, z_1) [35]. We will perform three successive rotations in the following manner. First we rotate φ degrees (see Fig. 23) as seen from a positive y axis about the z axis producing new axes z_1 and y_1. This produces the S_φ transformation on x, y, and z given by Eq. (57):

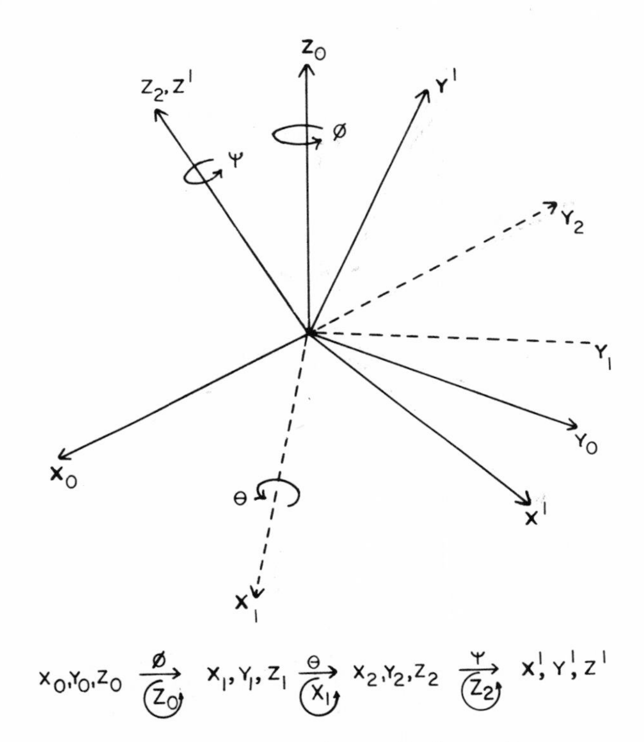

Fig. 23. Rotation of coordinate system.

$$S_{\varphi} = \begin{vmatrix} \cos\varphi & \sin\varphi & 0 \\ -\sin\varphi & \cos\varphi & 0 \\ 0 & 0 & 1 \end{vmatrix} \tag{57}$$

where

$$\begin{pmatrix} x_1 \\ y_1 \\ z_1 \end{pmatrix} = S_{\varphi} \begin{pmatrix} x \\ y \\ z \end{pmatrix} \quad \text{or}$$

$$x_1 = x \cos\varphi + y \sin\varphi + 0, \tag{58}$$

$$y_1 = -x \sin\varphi + y \cos\varphi + 0, \tag{59}$$

$$z_1 = 0 + 0 + z; \tag{60}$$

then a rotation θ around the x_1 axis, where

$$S_\theta = \begin{vmatrix} 1 & 0 & 0 \\ 0 & \cos\theta & \sin\theta \\ 0 & -\sin\theta & \cos\theta \end{vmatrix}, \tag{61}$$

and finally a rotation ψ around the x_2 axis where

$$S_\varphi = \begin{vmatrix} \cos\psi & \sin\psi & 0 \\ -\sin\psi & \cos\psi & 0 \\ 0 & 0 & 1 \end{vmatrix}. \tag{62}$$

The overall transformation $S = S_\varphi S_\theta S_\psi$ (which is not commutative) is given by

$$S = \begin{vmatrix} \cos\psi\cos\varphi - \cos\theta\sin\varphi\sin\psi & \cos\psi\sin\varphi - \cos\theta\cos\varphi\sin\psi & \sin\psi\sin\theta \\ -\sin\psi\cos\varphi - \cos\theta\sin\psi\cos\psi & -\sin\psi\sin\varphi + \cos\theta\cos\varphi\cos\psi & \cos\psi\sin\theta \\ [\sin\theta\sin\varphi & -\sin\theta\cos\varphi & \cos\theta] \end{vmatrix}. \tag{63}$$

[Note that (35) has an error in the S-matrix products.]

For projection on a two-dimensional surface the bracketed lower row of the matrix which represents the z coordinate can be omitted. If the figure is to rotate simultaneously around three angles then 14 multipliers are required. Such a sophisticated rotational capability is not necessary for display purposes. Usually one angle at a time will suffice; the other angles can be set to constant values, a procedure that converts a need for a multiplier to a simple need for a potentiometer. For example, suppose that we set $\theta = \tan^{-1}(1/\sqrt{2})$, $\psi = 0$, and rotate φ. Then S simplifies to

$$S = \begin{vmatrix} \cos\varphi & \sin\varphi & 0 \\ -\frac{\sqrt{6}}{3}\sin\varphi & \frac{\sqrt{6}}{3}\cos\varphi & \frac{\sqrt{3}}{3} \\ \frac{\sqrt{3}}{3}\sin\varphi & -\frac{\sqrt{3}}{3}\cos\varphi & \frac{\sqrt{6}}{3} \end{vmatrix}, \tag{64}$$

$$\bar{x} = x\cos\varphi + y\sin\varphi, \tag{65}$$

$$\bar{y} = -\frac{\sqrt{6}}{3}x\sin\varphi + \frac{\sqrt{6}}{3}y\cos\varphi. \tag{66}$$

The analog implementation of this is given in Fig. 24. In this $\sin\varphi$ and $\cos\varphi$ can either be generated from a function generator or, for continual rotations, more easily from the sin/cos generator described earlier. Figure 24 is representative of an analog circuit which makes a polar transformation instantaneously! A series of rotational positions for the sphere for which contours were derived earlier is shown in Fig. 25. Note that if a contour figure can be drawn without flicker then it can be rotated without flicker, i.e., real-time rotations are possible. Since the fidelity of a typical oscilloscope is only about

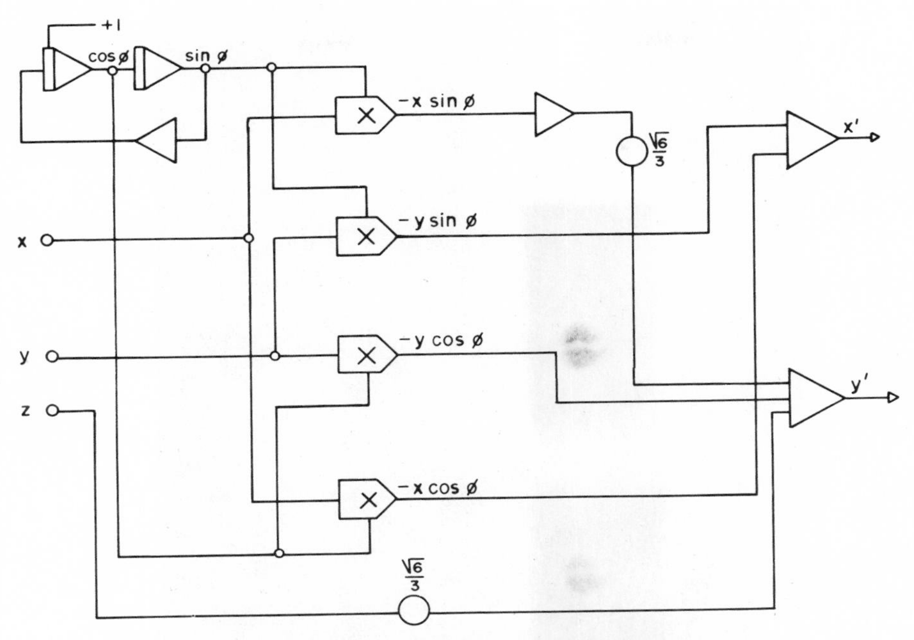

Fig. 24. Program for rotating figures.

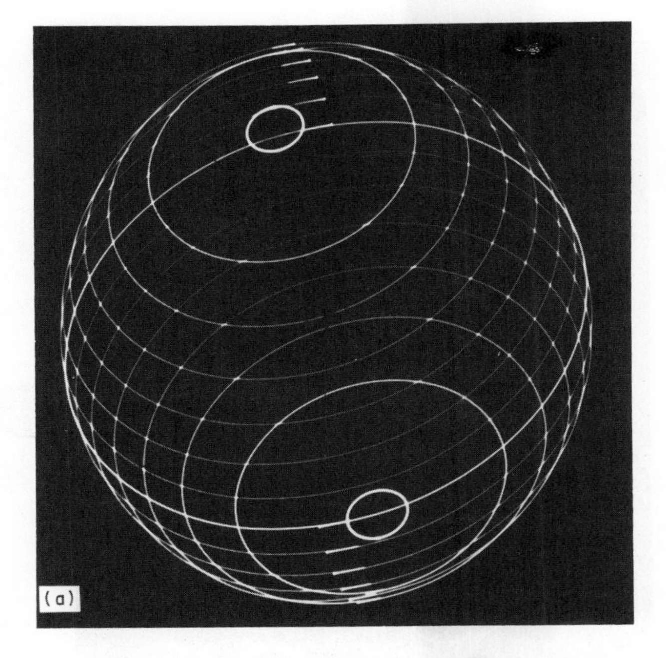

Fig. 25. Two-space projection of rotating sphere.

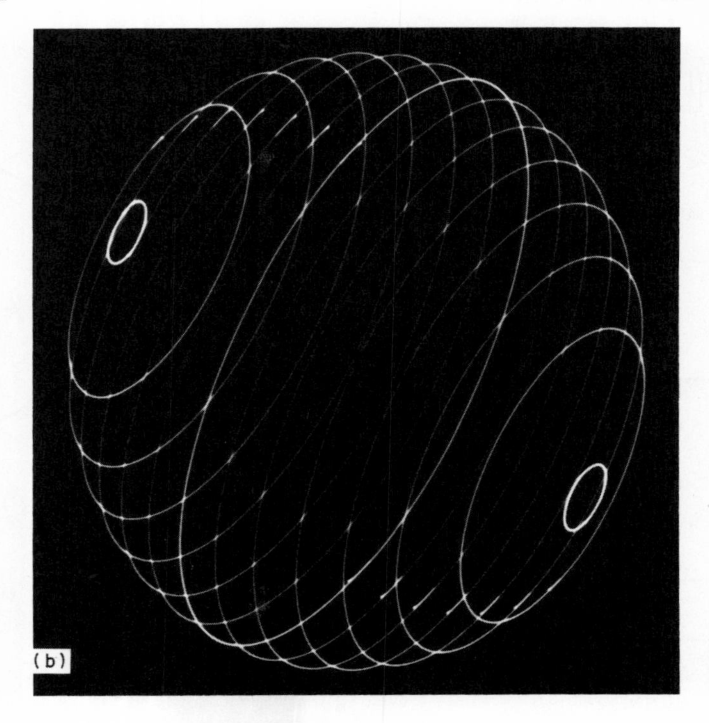

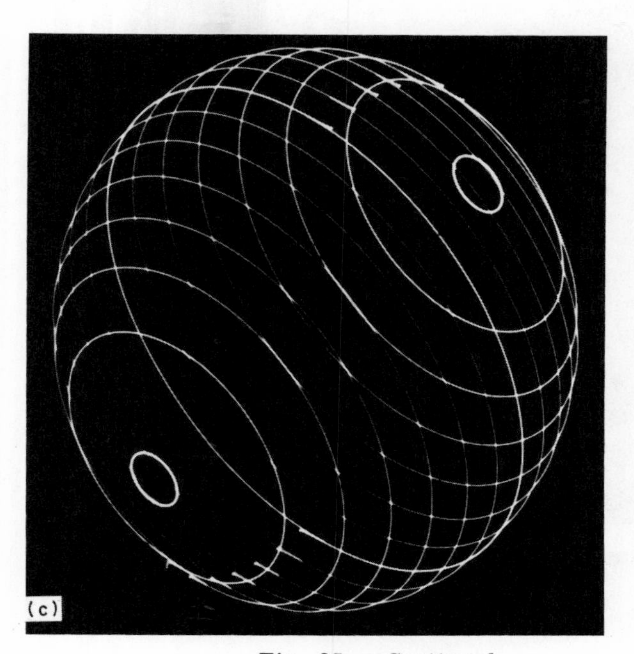

Fig. 25. — Continued.

1%, it is possible to compute contours at a rate of 0.1 msec each with an accuracy consistent with display. This would enable over 50 contour lines to be put in a figure and rotated without flicker.

IV. HYBRID COMPUTATION

A. General Introduction

Two things should now be clear. Logic control of an analog computer greatly magnifies its computational capability. However, with the limited number of logic elements on an analog/hybrid computer, the potential is barely being touched. Furthermore, from the logic and control features which have been demonstrated in the last examples it seems that the architecture of the analog/hybrid computer anticipates a higher level of control. Consider, for example, the overrides possible on all flip flops, comparators, counters, and mode control-overrides which can run, stop, load, and set these devices. That higher level is, of course, a digital computer.

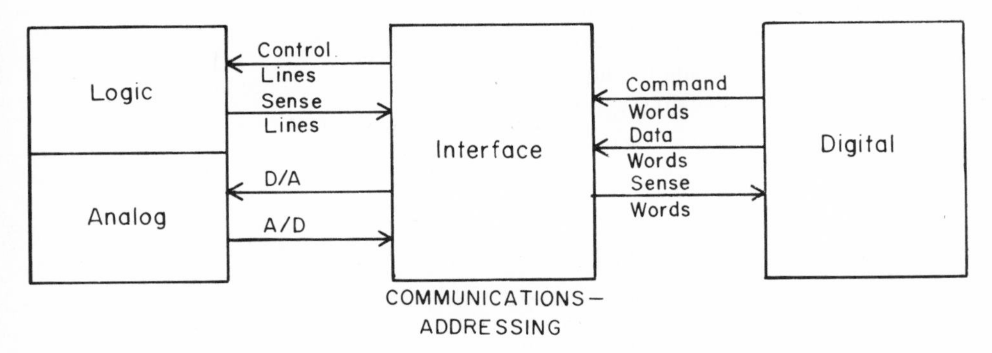

Fig. 26. Basic hybrid architecture.

The overall architecture for a typical hybrid is shown in Fig. 26. The bare minimum for a hybrid system in addition to the computers are Sense and Control Lines, an A/D converter, some D/A converters, and, most important, a communications link between them. The Sense and Control Lines (between eight and 16 each) terminate in addressable holes in the logic patch area and can be read and set by the digital computer. It is the Control Lines that can be patched to provide those overriding logic signals alluded to earlier. Since virtually all analog/hybrid computers have a digital voltmeter they also have an A/D converter. Of course, this A/D converter usually has a slow conversion rate (<100 conv/sec); but, though a bottleneck, it will do the job. A more complete hybrid system would have a high-speed A/D, as required for the earlier partial differential equation example. D/A converters will eventually be the coefficient device on all hybrid systems, replacing servo-set potentiometers. One would want at least 6 to 12 of these to begin with, preferably double-buffered. The communications link is the all important addressing link, enabling the digital computer to reach, control, and read as many as 500 components in the analog/hybrid computer.

At this point it would be useful to list a sample (Table 6) of the typical array of instructions available to control the analog computer. The specific ones tabulated are the machine or assembly language level instructions which accompany the Applied Dynamics AD/5 interface. That is, each control word in Table 6 corresponds to one 16-bit instruction. It is this instruction set which to a great measure will determine the power of the hybrid system.

TABLE 6

AD/5 Control Functions

Assembly Mnemonic	Operation
IC	Place AD/5 in Initial Condition Mode
OP	Place AD/5 in Operate Mode
H	Place AD/5 in Hold Mode
LD	Place AD/5 in Load Mode
RUN	Place AD/5 in Run Mode
STOP	Place AD/5 in Stop Mode
V1S	Place interval timer in 1 sec frame
V100M	Place interval timer in 100 msec frame
V10M	Place interval timer in 10 msec frame
X1	Place AD/5 in X1 time scale
X10	Place AD/5 in X10 time scale
X100	Place AD/5 in X100 time scale
X1000	Place AD/5 in X1000 time scale
SET COEF	Initiate a set coefficient operation
LST	Initiate a logic step
SP	Initiate a DRM[a] sample operation to digitalize an analog value

TABLE 6 (continued)

LE SET	Set Logic Exec state
LE CLR	Clear Logic Exec state
CLR	Clear the address or data register
SCLR	Clear the AD/5 system (initialize)
ADDR	Set the Address entry state
DATEN	Set the Data entry state
ENA 0	Enable LSD(BCD) of the DRM onto system read bus
ENA 1	Enable NMSD(BCD) of the DRM onto system read bus
ENA 2	Enable NMSD(BCD) of the DRM onto system read bus
ENA 3	Enable NMSD(BCD) of the DRM onto system read bus
ENA 4	Enable MSD of the DRM onto system read bus. Includes MSB, Sign, Busy, 1SP
ENA 5	Hybrid On
ENA 6	Hybrid Off

[a]Digital Ratio meter or Digital Voltmeter.

In addition to the AD/5 control functions in Table 6 there is a Control Register that contains 16 bits of information that terminate in 16 patch holes on the patch panel. The commands in Table 7 are available to it and, like those in Table 6, are one machine word instructions.

Finally, the Sense Register makes 16 bits of external data from the AD/5 available to the digital computer. The data word has the content shown in Table 8 with the commands given in Table 9.

The control commands to the analog/hybrid computer perform any operation that can be done manually at the control panel, e.g., setting/reading coefficient devices, setting the analog in IC, RUN, or HOLD, changing time scales, etc. The overriding control of a specific device (such as an integrator) which can be accomplished manually by a patch in the logic area can be accomplished under program control by committing a Control Line to that device and patching. It is apparent, then, that the digital computer can now

TABLE 7

Typical Control Register Commands[a]

DISABLE	communications
ENABLE	communications
ENABLE	next data communication only
CLEAR	register on ones
SET	register on ones
LOAD	data word
READ	register
CLEAR	register
DISABLE	synchronous operation
ENABLE	synchronous operation

[a]These are typical hybrid commands for the control of the status of 16 patch holes on the patch panel. These output patch holes represent the Control Register.

TABLE 8

Typical Sense Register Word[a]

Bits		
0-7	Interrupt 0-7	
8	Overload Status	
9	Hybrid on status	
10	Spare	
11	Spare	
12	Coefficient setting status	DRM data for least significant bit
13	DRM busy status	
14	Sign bit	
15	Most significant bit	

[a]Five commands are used through the control register to specify each of five sets of data to appear in Bits 12-15. The four less significant digits from the DRM are in BCD representation. The most significant digit word contains Bits 12-15 as indicated in the table.

TABLE 9

Typical Sense Register Commands

ENABLE	interrupt lines
DISABLE	interrupt lines
DISABLE	register inputs
ENABLE	(track) register inputs
STORE	(hold) register inputs
READ	register
CLEAR	register
LOUD	next data word

do anything that an operator could do, with the exception of patching up a problem. Before we consider the progress in the latter area, let us examine a typical control operation.

Suppose we want to set the potentiometer with address 123 to 0.4567. To do this by hand it is necessary to push 13 buttons in the proper order, as shown in Table 10. To accomplish the same task with a digital computer, the same sequence of instructions is required. This involves twice as many computer instructions, since each instruction must be loaded into a register in the digital computer before it is sent. If the device were a DAC, then the first digit in the device address would be a D (3). By substituting a SP instruction for the CLR instruction (8), substituting ENA X, DIB AD/5 for DOB AD/5 in the digital steps (9-13) the sequence in Table 10 would <u>read</u> the potentiometer (or amplifier, or multiplier, etc.).

The latter capability is quite useful in setting up a problem. For example, suppose a user has been "experimenting" with a simulation and wanted to continue, but had to leave for a while. With a hybrid system he can call a subroutine composed of iterations of Table 10 which would read every "settable" device in the main frame in less than a second. This could then be stored either in a protected file for the user or on paper tape. Upon returning, the user could quickly reset all devices to exactly their state at the time of departure.

The assembly or machine language level of hybrid operation is interesting as a way of understanding hybrid operation, but it is not very practical.

TABLE 10

A Sample Hybrid Operation: Calling the Address of a Coefficient Potentiometer and Setting It

Pushbutton steps	Manual[a] pushbutton operation	Digital[b] instruction	Operation	Display registers resulting from instructions	
				Address	Output or data
				D130	0.1469
		LDA 1, LA[c]	Enables receipt by AD/5 of control words		
		DOB[d] 1, AD/5	Command Sent		
				D130	0.1469
1	ADDR	LDA, ADDR	Enable Address Entry		
		DOA 1, AD/5	Data Sent		
				D130	0.1469
2	CLR	LDA 1, CLR	Clears Address Register		
		DOA 1, AD/5	Data Sent		
				A000	0.0000
3	P(1)	LDA 1, 1A	First digit in address register (A.R.)		
		DOA 1, AD/5	Data Sent		
				P000	0.0000
4	1	LDA 1, 2A	Second digit in A.R.		
		DOA 1, AD/5	Data Sent		
				P100	0.0000
5, 6	2, 3	Repeat (3A, 4A)	For next two digits		
				P123	0.0000
7	DATA ENTRY	LDA 1, DATEN	Enable Data Entry Register (DER)		
		DOA 1, AD/5	Data Sent		

				P123	0.7643
				(left in DER from last setting)	
8	DLR	LDA 1, CLR	Clear DER		
		DOA 1, AD/5	Data Sent		
				P123	0.0000
9	0	LDA 1, MSB	First (MSB) datum		
		DOA 1, AD/5	Data Sent		
				P123	0.0000
10	4	LDA 1, NMSB	Second datum		
		DOA 1, AD/5	Data Sent		
				P123	0.4000
11, 12, 13	5, 6, 7	Repeat for next 3 digits			
				P123	0.4567

[a] Applied Dynamics, AD/5.

[b] Data General, Nova Assembly Language.

[c] Each of the mnemonics loaded into the accumulator identifies a core location wherein the coded control or data word resides. Upon the execution of each LDA instruction that word is loaded into an accumulator ready for transmission to the interface.

[d] The B-register (DOB) in the interface controls entry to the specific data registers described in Tables 6, 7, 8, and 9. The A-register (DOA) accepts the data.

A higher level language is necessary. Most manufacturers of hybrid systems will provide such software with their systems. Thus, to be effective as a general purpose hybrid system we would want, for example, to be able to set a device with one instruction, such as SET P123, 0.4567, rather than with 26 instructions. A complete hybrid software package is an elaborate and powerful controller. Generally, it is composed of several basic parts which are identified in the flow diagram for a typical system shown in Fig. 27. The flow diagram assumes that the problem has been patched and all control features set. The program first does diagnostics and then runs the program.

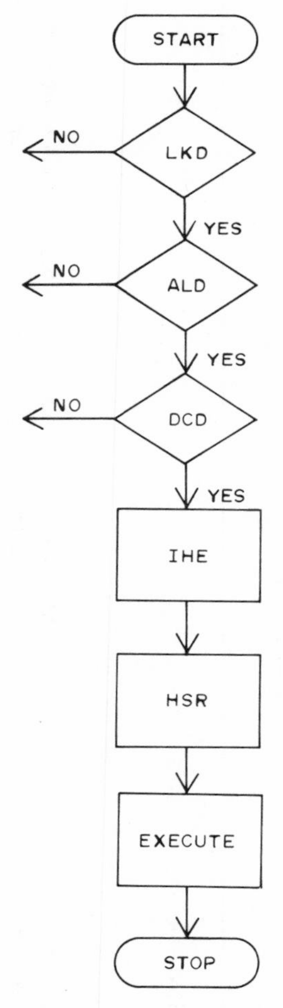

Fig. 27. Typical hybrid software flow diagram.

The first step in the program is a test of the interface linkage, for example, logic patches, checking all valid DACs, and sample/hold status, which are within the interface. The program then proceeds to a diagnosis of the logic and analog components on the computer. This is a most important check, often skipped in nonhybrid systems because of the tedium of this time-consuming chore. It is a natural for the digital computer. The ALD test is not run on the program logic/analog patch panel, but on a special diagnostic panel which allows all components to be exercised so that control, function, and reliability can be tested. It will repeat this for, say, 1000 times, and then at the end report that the system is ready to go or, if not, why. For example, amplifier A001 may have an offset, or perhaps the RUN mode failed x times. If the user wishes, he can then proceed to a digital computer exerciser to check it for problems. At this point the user has quickly and routinely satisfied himself that the system is quite operational — a check that he would probably never do without digital control, so that the hardware problems would only be discovered (if ever) by a series of tortuous deductions from symptoms in the solution of the problem. Now the problem patch panel is replaced and a hybrid executive takes over, which has at its disposal the hybrid subroutines for control and data transmission. After checking program errors, such as initial outputs from integrator and overloads, the problem is executed.

Hybrid software can become very elaborate with such possibilities as automatic scaling. The earlier languages were compilers usually using FORTRAN. More recently interpreter versions such as HYBASIC have become available.

B. Automatic Patching

There remains one remnant of the original analog computer even in the present state-of-the-art analog and hybrid systems which are commercially available today: patching. Patching is the time-consuming process which is probably most responsible for analog computers being as rare as they are. Recent developments in prototype autopatching systems are worth mentioning at this point since they are harbingers of the future for hybrid computation.

Theoretically an automatic patching system should be possible by simply constructing a switching matrix that would enable the output of any device to be connected to the input of any device. Since in most analog programs two outputs and one input are used on the average for each device, if the number of analog devices is n, the total number of switches required with this strategy is $2n^2$. Clearly, even with simple programming requirements, hundreds of switches are required. Until recently the power and space

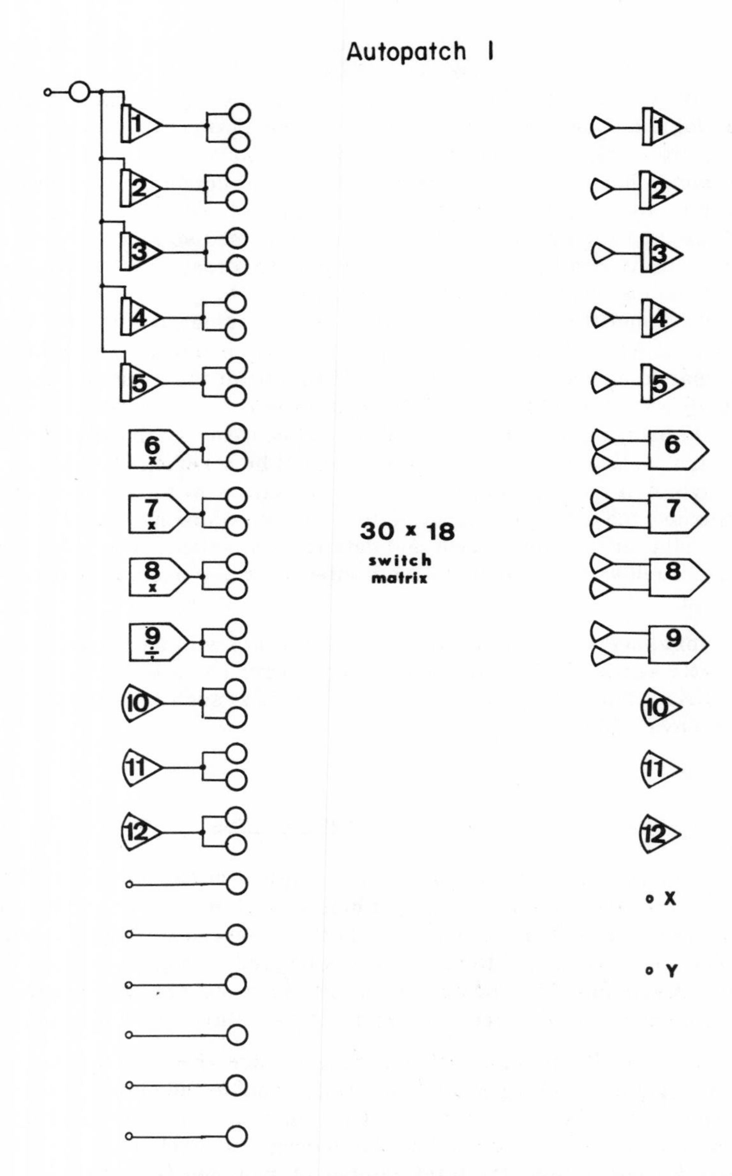

Fig. 28. A direct connection switch matrix for autopatch.

demands for such an array of switches was prohibitive. Now with the availability of small, lower power FET switches the picture has changed. Accordingly, several approaches to automating patching systems have been explored [39, 40] and five prototypes have been built [41-44, 48].

TABLE 11

Switch Matrix Printout for Autopatched $\ddot{z} = -kz^a$

		Outputs																
		I1		I2		I3		I4		I5		I6		S1		S2		
Address	Inputs	0	1	2	3	4	5	6	7	8	9	10	11	12	13	14	15	← Bit number
7242	I1			P														
7243	I2	P																
7244	I3																	
7245	I4																	
7246	I5																	
7247	I6																	
7250	S1																	
7251	S2																	
7252	M1																	
7253	M1																	
7254	M2																	
7255	M2																	
7256	M3																	
7257	M3																	
7260	D1																	
7261	D2																	
7262	X_{out}					D												
7263	Y_{out}		D															

TABLE 11 (continued)

Address	Inputs	Outputs: M1 (bits 0 1)	M2 (bits 2 3)	M3 (bits 4 5)	D1 (bits 6 7)	F1	F2	F3	F4	F5	F6
7264	I1										
7265	I2										
7266	I3										
7267	I4										
7270	I5										
7271	I6										
7272	S1										
7273	S2										
7274	M1										
7275	M1										
7276	M2										
7277	M2										
7300	M3										
7301	M3										
7302	D1										
7303	D2										
7304	X_{out}										
7305	Y_{out}										

← Bit number

[a] I = Integrator, S = Summer, M = Multiplier, D = Divider, F = Special function. The P-D entries in the above matrix represent closed switches for patch and display program shown below. All other switches are open, i.e., "0".

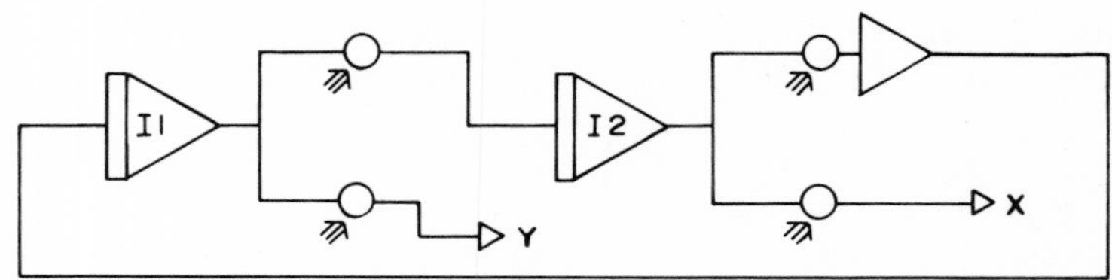

Basically, two strategies exist: direct, switchable connections to everything, or trunks with intermediate switch points. An example of the first type is shown in Fig. 28. For a relatively small number of analog components requiring fewer than 1000 switches this method is relatively simple to construct. For programs requiring more components a configuration consisting of several clusters of components directly connected within each cluster but with trunk connections between clusters is probably the most efficient system. The 540 switches required for the system shown in Fig. 28 are individually addressable and readable. For example, suppose we wanted to solve the classic second-order differential equation $\ddot{z} = -kz$; closing switches shown by P's in Table 11 would do it. If we want to display $\dot{z}$ on the y (vertical) axis and z on the x (horizontal) axis, closing the switches shown by D's in Table 11 would accomplish that. Patch buffers for the switch matrix for that solution and others are shown in Table 12.

TABLE 12

Examples of Patch Buffers

Equation	Display: X axis	Display: Y axis	Patch buffer addresses with nonzero contents
$\ddot{z} = -kz$	z	$\dot{z}$	7242, 020000; 7243, 100000; 7262, 010000; 7263, 040000
$\dot{x} = +k_1x - k_2xy$ Prey $\dot{y} = -k_3y + k_4xy$ Predator	x	y	7242, 100010; 7243, 020000; 7252, 050000; 7254, 050000; 7262, 100000; 7263, 020000; 7264, 020000; 7272, 100000
$\ddot{\psi} = k_1t^2\psi - k_2\psi$	t	ψ	7242, 020000; 7243, 100000; 7245, 004000; 7252, 010000; 7256, 001000; 7262, 002000; 7263, 010000; 7264, 100000

Initial conditions can be set with one DAC by using the Control Lines to set all integrators to HOLD except the one to be initialized, initialize that one, set to HOLD, and then shift to the next integrator, repeating, and so on. In addition to the attractive features ofsize and power consumption, FET switches are also fast. Furthermore, the switches can be doubly buffered, i.e., a controlling memory and a back-up memory for each switch. At this point all of the players have been introduced for the following action. The high-speed operation of the analog, including high-speed doubly buffered coefficient devices coupled with a high-speed doubly buffered switching matrix, makes it

TABLE 13

Timing Sequence for Autopatch

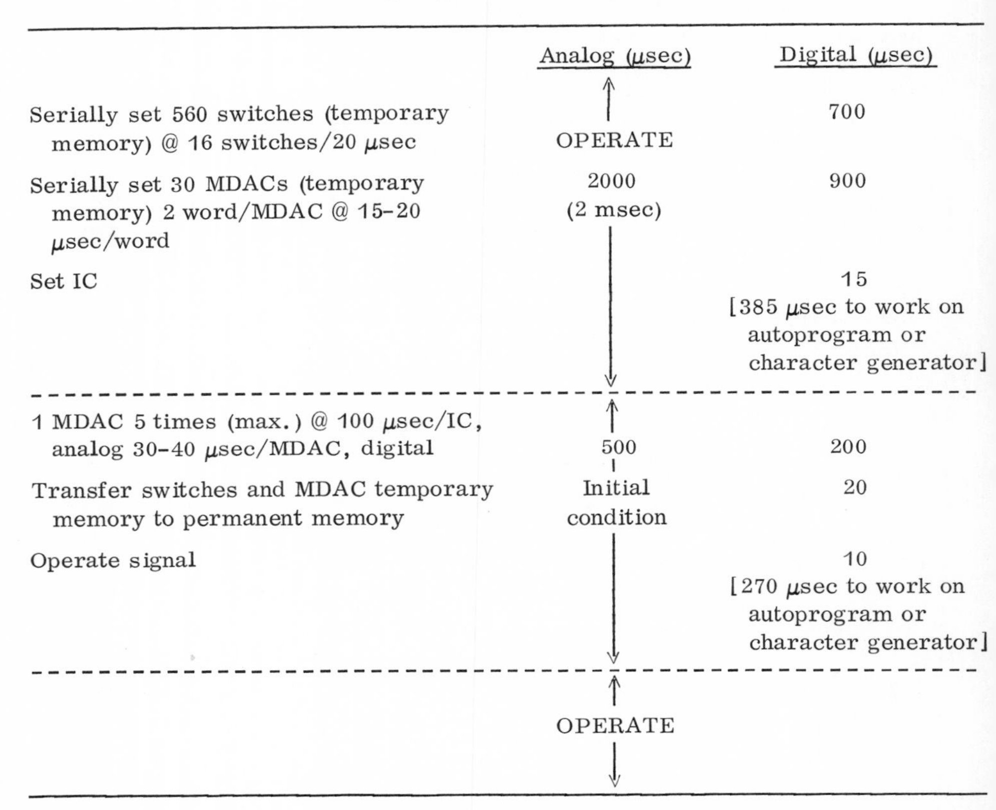

	Analog (μsec)	Digital (μsec)
Serially set 560 switches (temporary memory) @ 16 switches/20 μsec	↑ OPERATE	700
Serially set 30 MDACs (temporary memory) 2 word/MDAC @ 15-20 μsec/word	2000 (2 msec)	900
Set IC	↓	15 [385 μsec to work on autoprogram or character generator]
1 MDAC 5 times (max.) @ 100 μsec/IC, analog 30-40 μsec/MDAC, digital	↑ 500	200
Transfer switches and MDAC temporary memory to permanent memory	Initial condition	20
Operate signal	↓	10 [270 μsec to work on autoprogram or character generator]
	↑ OPERATE ↓	

possible to solve a totally different differential equation with totally different coefficients every 2.5 msec. The timing sequence for an operational version of the system in Fig. 28 is shown in Table 13. The doubly buffered DAC and switches enables the digital computer to load the next problem consisting of switch settings and DAC settings into the temporary buffer while the current problem is being solved and plotted by the analog computer. When the analog is no longer busy a single strobe command will load the next problem in less than 10 μsec.

But why would someone want to look at the solution of a totally different differential equation every 2.5 msec? One reason, and the primary goal for the system just described, is time-sharing. Having gotten this far it would be a relatively simple matter to multiplex several terminals to the hybrid system. Then each terminal user could set up his own problem and interact with it. With a delay of $m(2.5 \times 10^{-3})$ sec, where m is the number of users, ten users could use the system independently with flicker-free display capability. More users could be accommodated with storage oscilloscopes or terminals with some memory capability. The system would permit real-time interactive displays at a reasonable cost.

C. Automatic Programming

The potential of autopatching is great, but it can be enormously strengthened with an autoprogramming software. Because of the rather mechanical nature of the general method of programming, it seemed possible to design an algorithm on the digital computer to do it automatically. Thus, the system could accept a statement of a differential equation with appropriate parameters and convert that into switch and MDAC settings. The following is a description of an operational autoprogramming system [43]. Other autoprogramming systems have been proposed [46, 47, 49].

Autoprogram is divided into eight parts. They are:

AUTOPROGRAM — Converts statement of problem into switch. DAC, and IC settings.

SYSTINT — Central branching point for all interrupts.

ASBI — Converts ASCII to binary (fixed point for DACs).

ICOEFF — Accepts coefficients from keyboard.

CONTROL — Controls who gets ICOEFF.

AD5INT — Senses interrupts from AD/5.

IOUT — Output of characters to terminals.

IOIN — Keyboard inputs.

Autoprogramming easily falls within the memory capacity restraint (8K word) of a modest minicomputer system. Minimal I/0 buffers totalling only 200 words are the only requirement for each terminal, plus 2800 words for Autoprogram itself. Thus, autopatch-autoprogram for up to 16 terminals can be handled with 8K 16-bit word minicomputer with over 300 words for each buffer. Furthermore, for all of the second-order differential equations attempted so far, Autoprogram appears to take less than 0.1 sec. After Autoprogram has been run once on a problem for a terminal user, it need not be rerun so long as the user sticks with that problem. In summary then, Autoprogram is fast and economical in core.

Autoprogram operates in the following manner. Suppose that Control is having Autoprogram work on someone's problem when someone at another terminal depresses a keyboard key. This causes a Control Interrupt due to a character coming in. This is the highest priority, so Control pauses while the character is entered into the 100-word Character buffer allocated for that terminal; then Control resumes its job. When the character stream for entering the problem by the user is complete, he depresses a control key followed by A. The next time that terminal is serviced the characters in the Character buffer will undergo an Autoprogram. When the Autoprogram is completed the Character buffer now contains the autopatch switch settings and the buffer is transferred to a 100-word Plot buffer for that terminal. Control then requests the terminal user to assign values for each coefficient identified in the problem. These inputs are identified by the control key followed by C and are entered in the Character buffer. When Coefficient sees that all coefficients have been given a value it stops asking, transfers the buffer to the Plot buffer, and awaits a run command. The user initiates this by depressing the control key followed by R. The next time Control is interrupted by the AD/5 signifying that it is ready and it is the turn for this terminal, the Plot buffer will be transferred to the AD/5 and then a command is given to computer. The run command can also cause Reset to clear both buffers for that terminal.

With the use of an analog character generator a truly interactive display system for simulations is now possible which can be used by a student or any user with no programming experience. This system is a prototype, and considerably more effort is required to obtain a reliable system with an expanded array of analog components under Autopatch control. But the system is noteworthy because it has demonstrated for the first time that a fully automatic time shared hybrid system is possible at moderate costs. Perhaps the next time that a review of hybrid computation is written, autopatching and autoprogramming will be commercially available.

APPENDIX: UNIFORM GRAPHICS FOR SIMULATION [2]

The symbols and the methods of laying out analog and hybrid computer diagrams presented here are advocated to alleviate the confusion caused by the uncoordinated invention of new symbols and diagramming practices. The increasing use of hybrid techniques and equipment has aggravated an already bad situation to the point that it is often no longer possible for one worker in our field to read another's diagram. Usually this is because symbols are devised and diagrams are drawn to include details peculiar to a particular kind of equipment. Such a wiring, or "patching," diagram is of course necessary for setting up and checking out an actual simulation, but hardware-peculiar details are only confusing to those with other kinds of equipment. With few exceptions, the use of a simplified signal-flow diagram to illustrate technical articles is much more effective.

With the foregoing in mind an SCi committee composed of George Burgin, Joe Hussey, Hans Jorgensen, Granino Korn, and John McLeod, selected the symbols and offers the following suggestions for their use. Primary considerations were current usage, clarity, and simplicity. We devised no new symbols and, unless there were overriding indications to the contrary, we adopted those already in widest use. Clarity and simplicity, we believe, will be enhanced by the choice of unique shapes to represent different components, and the elimination of all unnecessary details in diagrams.

There was no intent on the part of our committee to set up standards for the industry. However, all diagrams appearing in Simulation will be prepared according to the committee's recommendations (as they may be modified from time to time), and we hope that these recommendations will prove attractive to others. Suggestions for modifications and additions are solicited.

General Rules

The following methods and symbols are recommended for the illustration of technical articles prepared for publication. Unless the purpose of the article is to describe the use of a particular kind of equipment, and the hardware details are pertinent to the subject, such illustrations should not be "hardware-peculiar." In other words, the objective should be to show signal flow, rather than "patching" details.

The primary, or overall, system diagram should show only the essential signal flow. Where it is necessary to show details, separate diagrams should be made and referenced to the primary diagram by enclosing the detailed area of the primary diagram in dotted lines with suitable notation.

The direction of signal flow should be indicated by arrowheads except where the shape of the symbols makes the direction of flow obvious. Primary signal flow (with the exception of feedback loops) should be from left to right, and, if practical, each "line" of symbols should be made to read like the mathematical relation it represents.

The choice of whether to end a line and label it (preferably with the symbol of the variable that the signal represents) when it reaches the right-hand side of the diagram, and then indicate its continuation with the same label as it enters again at the left-hand side, instead of drawing in the connection, should be made on the basis of clarity; if a line returning the signal from right to left will cross many other lines and be hard to follow, it should not be drawn in.

If a diagram involves a number of identical circuits, only one should be shown in detail, while the others should be indicated by boxes with appropriate notation.

Components should not be numbered unless they are referred to by number in the text.

All amplifier gains should be shown just outside the amplifier at the point where the input enters.

Always apply the test of clarity and simplicity. Ask yourself: "Is this the most understandable way to diagram this for those unfamiliar with the hardware, and less familiar with the subject, than I?"

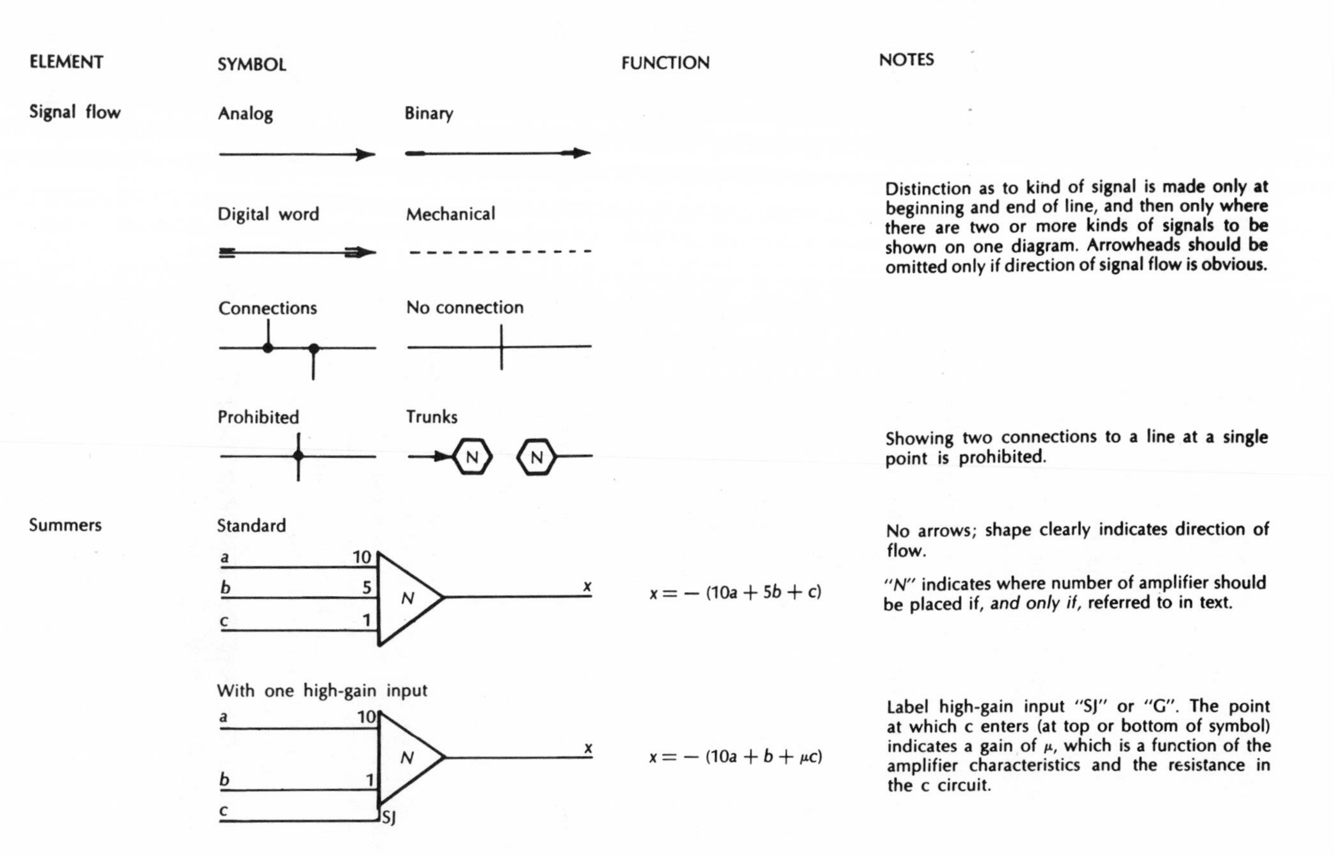

ELEMENT	SYMBOL		FUNCTION	NOTES
Signal flow	Analog	Binary		
	Digital word	Mechanical		Distinction as to kind of signal is made only at beginning and end of line, and then only where there are two or more kinds of signals to be shown on one diagram. Arrowheads should be omitted only if direction of signal flow is obvious.
	Connections	No connection		
	Prohibited	Trunks		Showing two connections to a line at a single point is prohibited.
Summers	Standard		$x = -(10a + 5b + c)$	No arrows; shape clearly indicates direction of flow. "N" indicates where number of amplifier should be placed if, *and only if*, referred to in text.
	With one high-gain input		$x = -(10a + b + \mu c)$	Label high-gain input "SJ" or "G". The point at which c enters (at top or bottom of symbol) indicates a gain of μ, which is a function of the amplifier characteristics and the resistance in the c circuit.

High-gain

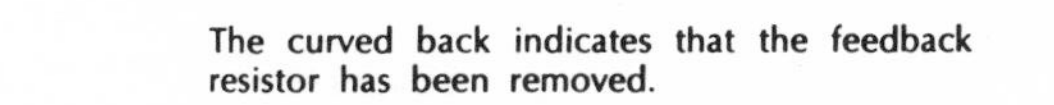

The curved back indicates that the feedback resistor has been removed.

Inverting only

$x = -a$

When amplifiers are used *only* to change signal sign they should be smaller. Orientation should be that which makes for greatest graphical simplicity, *i.e.*, they may be shown in vertical, or in feedback, paths.

Integrators

Standard

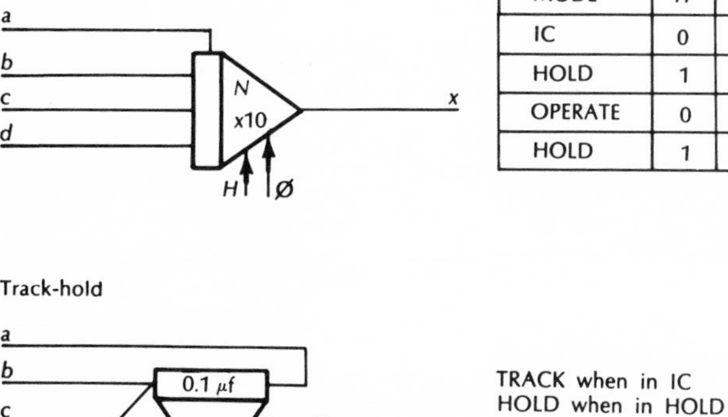

$$x = -a - \int_0^t (10b + 5c + d)\, dt$$

Standard integrator modes and time scales are those of the basic problem. If they are *not*, the facts should be noted as indicated in the diagram below.

Three-mode

MODE	H	Ø
IC	0	0
HOLD	1	0
OPERATE	0	1
HOLD	1	1

Arrows indicating mode control should be shown only when two or more integrators are operating in different modes simultaneously. The mode-control inputs should then be labeled to conform to a truth-table. Because the truth-table will be equipment-peculiar it should be shown somewhere on each patching diagram to which it applies. A typical one is shown.

The integrator time-scale, when other than that of the basic problem, should also be as indicated. The amplifier illustrated is integrating at ten times problem rate.

Track-hold

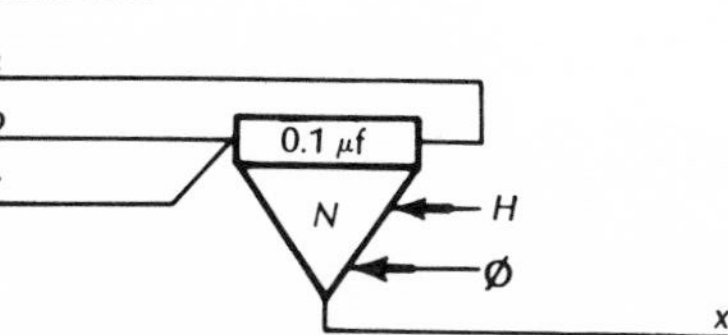

TRACK when in IC
HOLD when in HOLD
(See truth-table)

When used in this way the size of the integrating capacitor may be important, as it affects the tracking lag. In such cases the value should be indicated as shown. The quantity tracked or held is $-(a + b + c)$.

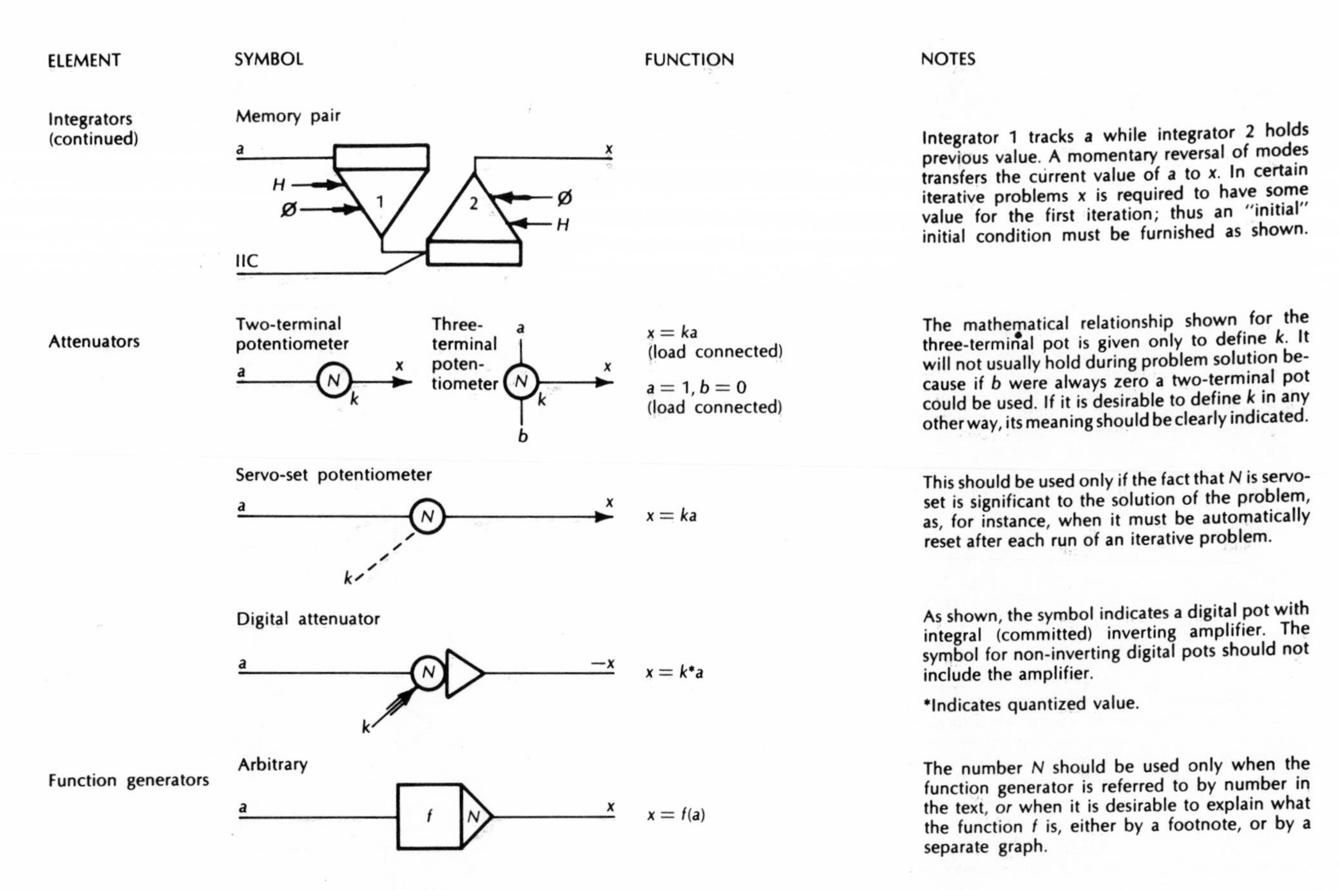

ELEMENT	SYMBOL	FUNCTION	NOTES
Integrators (continued)	Memory pair		Integrator 1 tracks a while integrator 2 holds previous value. A momentary reversal of modes transfers the current value of a to x. In certain iterative problems x is required to have some value for the first iteration; thus an "initial" initial condition must be furnished as shown.
Attenuators	Two-terminal potentiometer; Three-terminal potentiometer	$x = ka$ (load connected) $a = 1, b = 0$ (load connected)	The mathematical relationship shown for the three-terminal pot is given only to define k. It will not usually hold during problem solution because if b were always zero a two-terminal pot could be used. If it is desirable to define k in any other way, its meaning should be clearly indicated.
	Servo-set potentiometer	$x = ka$	This should be used only if the fact that N is servo-set is significant to the solution of the problem, as, for instance, when it must be automatically reset after each run of an iterative problem.
	Digital attenuator	$x = k^{*}a$	As shown, the symbol indicates a digital pot with integral (committed) inverting amplifier. The symbol for non-inverting digital pots should not include the amplifier. *Indicates quantized value.
Function generators	Arbitrary	$x = f(a)$	The number N should be used only when the function generator is referred to by number in the text, or when it is desirable to explain what the function f is, either by a footnote, or by a separate graph.

Mathematical functions (typical)

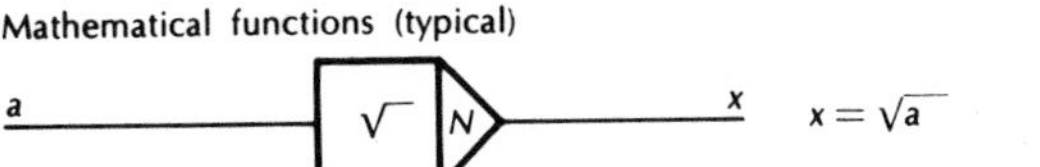

$x = \sqrt{a}$

In cases where the function generated can be represented by a standard mathematical symbol, the *f* should be replaced by the symbol.

Multiplier

α
+a
+b
N_i
−x

$x = \frac{ab}{\alpha}$

$x = \frac{\alpha a}{b}$

These symbols should be used for all analog multipliers and dividers; if the *kind* is significant to the solution of the problem, it should be so stated in the text and/or noted on the diagram. Because equipment differs, the sign of the output for specified signs of both inputs should be given, otherwise inversion will be assumed.

The number *N* should be used only if the component is referenced in the text, *or* if more than one channel of a multichannel device is used. In the latter case the subscript, in this case *i*, identifies the channel. In many cases, it will be found that a drawing can be simplified and clarified by drawing the same multichannel device in more than one signal flow path. In this case the number would be the same, but the subscript would be different for each channel.

Divider

α
a
b
N_i
−x

α = some reference voltage, not necessarily the computer reference voltage

Resolvers

Polar to rectangular

R
θ
N
x
y

Rectangular to polar

x
y
N
R
θ

$x = R \cos \theta$
$y = R \sin \theta$

$\theta = \arctan y/x$
$R = \sqrt{x^2 + y^2}$

If a resolver has additional outputs they should be shown only if used.

Comparators

With binary output

a
b
N
U

With true and false binary outputs

a
b
N
U
$\tilde{U}$

$U = 1 \quad (a + b) > 0$
$U = 0 \quad (a + b) < 0$
$\tilde{U} \neq U$

The symbol for a relay comparator can, of course, be made by combining either of these symbols with that of a relay.

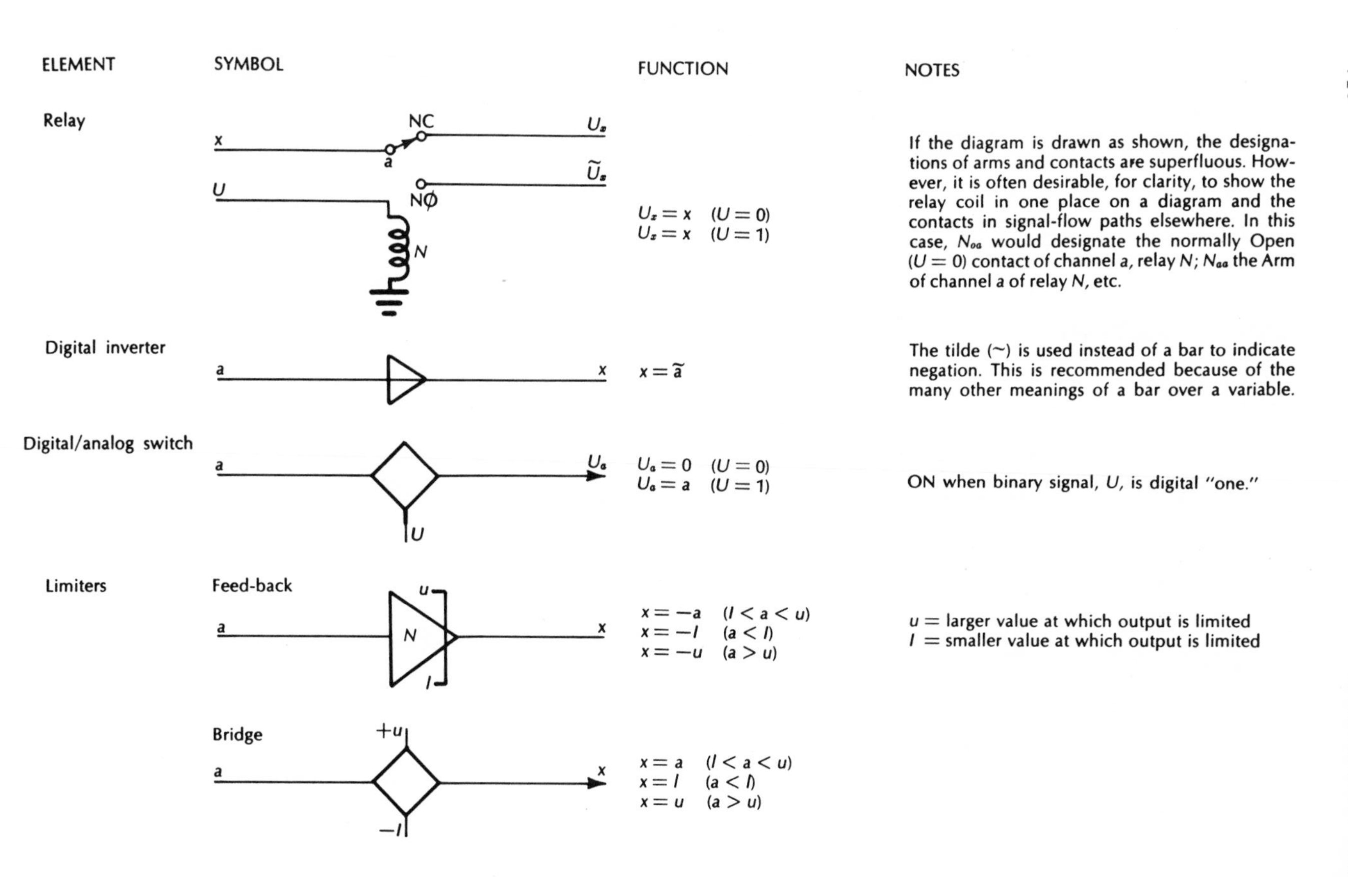

ELEMENT	SYMBOL	FUNCTION	NOTES
Relay	x, a, NC, U_x, U, NØ, $\tilde{U}_x$, N	$U_x = x \quad (U = 0)$ $U_x = x \quad (U = 1)$	If the diagram is drawn as shown, the designations of arms and contacts are superfluous. However, it is often desirable, for clarity, to show the relay coil in one place on a diagram and the contacts in signal-flow paths elsewhere. In this case, N_{oa} would designate the normally Open ($U = 0$) contact of channel a, relay N; N_{aa} the Arm of channel a of relay N, etc.
Digital inverter	a, x	$x = \tilde{a}$	The tilde (~) is used instead of a bar to indicate negation. This is recommended because of the many other meanings of a bar over a variable.
Digital/analog switch	a, U_a, U	$U_a = 0 \quad (U = 0)$ $U_a = a \quad (U = 1)$	ON when binary signal, U, is digital "one."
Limiters	Feed-back a, u, N, l, x	$x = -a \quad (l < a < u)$ $x = -l \quad (a < l)$ $x = -u \quad (a > u)$	u = larger value at which output is limited l = smaller value at which output is limited
	Bridge a, $+u$, $-l$, x	$x = a \quad (l < a < u)$ $x = l \quad (a < l)$ $x = u \quad (a > u)$	

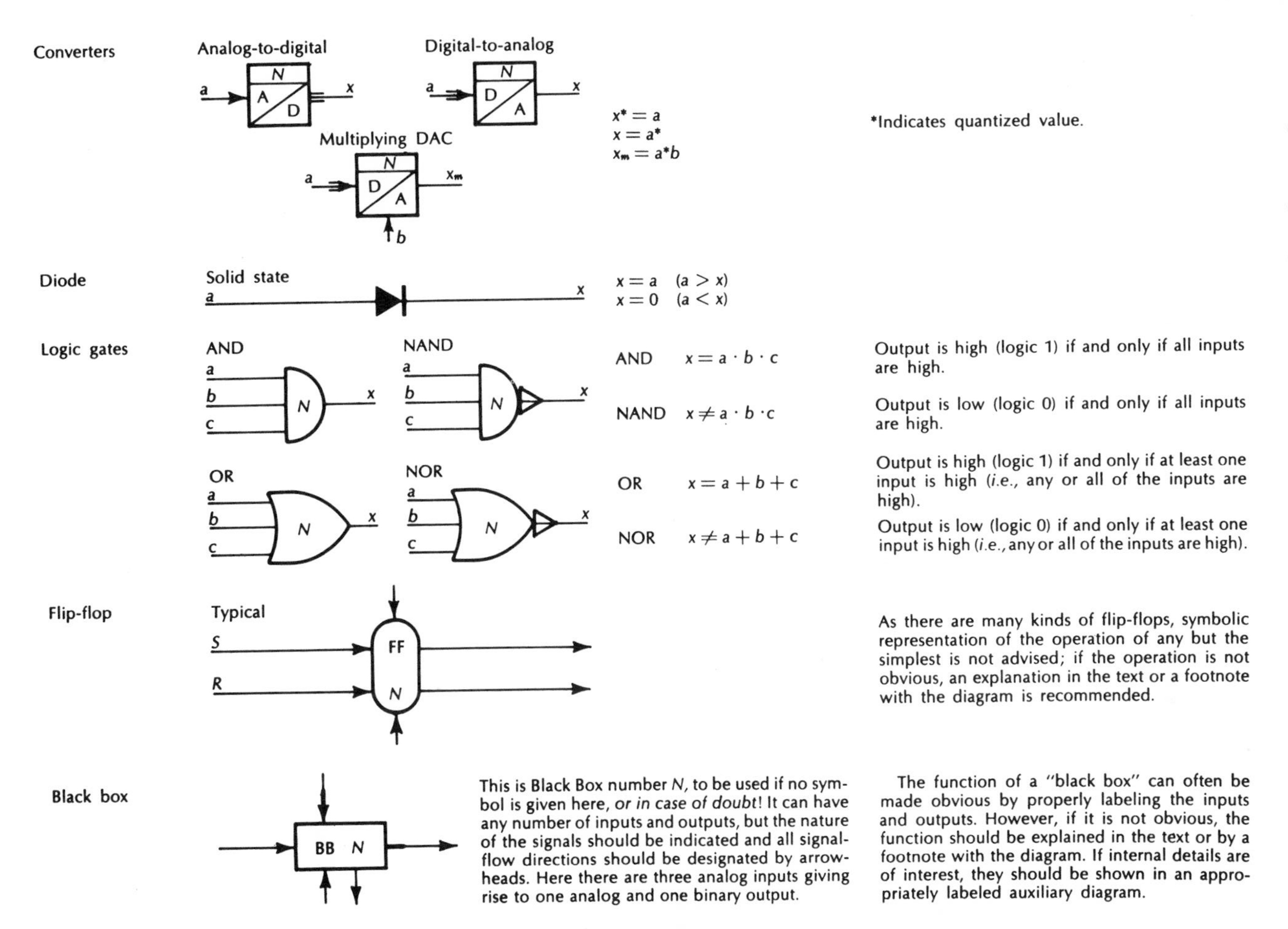

	Symbol	Function	Notes
Converters	Analog-to-digital; Digital-to-analog; Multiplying DAC	$x^* = a$ $x = a^*$ $x_m = a^*b$	*Indicates quantized value.
Diode	Solid state	$x = a \quad (a > x)$ $x = 0 \quad (a < x)$	
Logic gates	AND	AND $x = a \cdot b \cdot c$	Output is high (logic 1) if and only if all inputs are high.
	NAND	NAND $x \neq a \cdot b \cdot c$	Output is low (logic 0) if and only if all inputs are high.
	OR	OR $x = a + b + c$	Output is high (logic 1) if and only if at least one input is high (*i.e.*, any or all of the inputs are high).
	NOR	NOR $x \neq a + b + c$	Output is low (logic 0) if and only if at least one input is high (*i.e.*, any or all of the inputs are high).
Flip-flop	Typical		As there are many kinds of flip-flops, symbolic representation of the operation of any but the simplest is not advised; if the operation is not obvious, an explanation in the text or a footnote with the diagram is recommended.
Black box		This is Black Box number *N*, to be used if no symbol is given here, *or in case of doubt*! It can have any number of inputs and outputs, but the nature of the signals should be indicated and all signal-flow directions should be designated by arrowheads. Here there are three analog inputs giving rise to one analog and one binary output.	The function of a "black box" can often be made obvious by properly labeling the inputs and outputs. However, if it is not obvious, the function should be explained in the text or by a footnote with the diagram. If internal details are of interest, they should be shown in an appropriately labeled auxiliary diagram.

REFERENCES

1. J. Clancy, Simulation, 8, 19 (1967).
2. G. Burgin, J. Hussey, H. Jorgensen, G. Korn, and J. McLeod, Simulation, 9, 275 (1967).
3. G. Hannauer, Handbook of Analog Computation, EAI West Long Branch, N.J., 1966.
4. G. Korn and T. Korn, Electronic Analog and Hybrid Computers, 2nd ed., McGraw-Hill, 1972, p. 16.
5. D. Rummer, Introduction to Analog Computer Programming, Holt, Rinehart & Winston, New York, 1969.
6. A. Jackson, Analog Computation, McGraw-Hill, New York, 1960.
7. J. J. Blum, Introduction to Analog Computation, Harcourt, Brace and World, New York, 1969.
8. J. R. Ashley, Introduction to Analog Computation, Wiley, New York, 1957.
9. C. L. Johnson, Analog Computer Techniques, McGraw-Hill, New York, 1956.
10. R. M. Howe, Introduction to Analog Computation Applied Dynamics, Ann Arbor, Michigan.
11. (a) L. Levine, Methods for Solving Engineering Problems, McGraw-Hill, New York, 1964; (b) p. 52; (c) p. 56.
12. Thomson, W. (Lord Kelvin), Proc. Roy. Soc., 24, 271 (1876).
13. D. L. Shirer, Applications Notes, ACEUG, Los Angeles II, 3.
14. M. L. Corrin, J. Chem. Ed., 43, 576 (1966).
15. F. D. Tabbutt, J. Chem. Ed., 44, 64, 486 (1967).
16. A. Hausner, IRETEC, Feb. (1962).
17. A. J. Lotka, JACS, 42, 1595 (1920).
18. J. Higgins, Industrial and Engineering Chemistry, 59, 19 (1967).
19. G. A. Bekey and W. J. Karplus, Hybrid Computation, Wiley, New York, 1968.
20. S. K. Hsu and R. M. Howe, Proc. FJCC, 1968, p. 601.
21. R. Vichnevetsky and K. A. Bishop, "Proceedings of a Seminar on Computer Methods for the Solution of Partial Differential Equations Applied to Environmental and Natural Resource Problems," Electronic Associates, West Long Branch, 1971.

22. R. M. Howe, "The DCU," Applied Dynamics, Ann Arbor, 1969.

23. G. S. Stubbs, "Application of Time-Sharing Techniques to Simulate Dynamic Processes Associated with Fluid Stream," EAI Hybrid Computer Course, 1962.

24. T. Miura and J. Iwata, Ann. AICA, 5, 141 (1963).

25. E. E. L. Mitchell, EAI SAG Report 19, Oct. 25, 1963.

26. I. M. Kolthoff and J. J. Lingane, Polarography, 2nd ed., Wiley-Interscience, New York, 1952, pp. 18-25.

27. L. I. Grossweiner, W. D. Brannan, and A. G. Vacroux, The Application of Analog Computers in Chemical Research, IIT, Chicago, 1969.

28. J. E. Stice and B. S. Swanson, Electronic Analog Primer, Blaisdell, New York, 1965.

29. H. V. Houston, Principles of Mathematical Physics, McGraw-Hill, New York, 1948, p. 5.

30. R. D. Sacks and H. B. Mark, Simplified Circuit Analysis, Dekker, New York, 1972.

31. H. V. Molmstadt and C. G. Enke, Digital Electronics for Scientists, Benjamin, New York, 1969.

32. Basics of Parallel, Hybrid Computers, Electronic Associates, Long Branch, N.J., 1967.

33. (a) AD/5 Reference Manual, Applied Dynamics, Ann Arbor, Michigan, 1971, p. 4(31); (b) pp. 591-621; (c) p. 5(2-7); (d) p. 4(35).

34. T. Newton, "On Using the Electronic Analog Computer in Teaching Calculus and Differential Equations," First Annual Joint Computer Conference, Iowa State University, 1970.

35. H. J. White and S. Tauber, Systems Analysis, Sanders, Philadelphia, Pennsylvania, 1969, pp. 42-43.

36. J. H. Lukes, Trans. IEEE Electronic Computers, EC-16, 133 (1967).

37. Von Selahattin, Technische Mitteilungen PTT, 9, 342 (1950).

38. O. H. Schmitt, J. Appl. Phys., 18, 819 (1947).

39. D. A. Starr and J. J. Jonsson, Simulation, 10, 281 (1968).

40. T. J. Gracon and J. C. Strauss, Simulation, 13, 133 (1969).

41. G. Hannauer, Simulation, 12, 219 (1969).

42. R. M. Howe, R. A. Moran, and T. D. Burge, Simulation, 15, 105 (1970).

43. (a) F. D. Tabbutt, "A Hybrid Approach to Computer Assisted Instruction," Proceedings of Conference on Computers in Chemical Education and Research, July 19-23, NIU, Dekalb, Illinois, 1971. (b) F. D. Tabbutt, "An Autopatch-Autoprogram Instructional Hybrid System," Proceedings of the Summer Simulation Conference, June 13-16, San Diego, California, 1972.

44. A. Yagi, Simulation, 17, 80 (1971).

45. R. Griswold and J. F. Haugh, J. Chem. Ed., 45, 576 (1968).

46. C. Green, IRETEC, October 1962.

47. M. L. Stein, IEEETEC, April 1963.

48. J. F. Shoup and W. S. Adams, Simulation, 18, 142 (1972).

49. H. B. Rigas and D. J. Coombs, Simulation, 19, 133 (1972). (b) H. B. Rigas and D. J. Coombs, IEEE Transactions on Computers, 1140 (1971).

Chapter 4

ON-LINE CLASSROOM COMPUTING IN CHEMISTRY EDUCATION VIA VIDEO PROJECTION OF TELETYPE OUTPUT

Ronald W. Collins

Department of Chemistry
Eastern Michigan University
Ypsilanti, Michigan

I. INTRODUCTION TO ON-LINE CLASSROOM COMPUTING AS A METHOD FOR COMPUTER-AIDED INSTRUCTION IN CHEMISTRY

The use of computers for instructional purposes in the chemistry curriculum normally involves a one-to-one student-computer interaction. This communication can be effected either via batch processing using card input, or via a remote teletype terminal used in time-shared mode. The role of the computer can range from that of surrogate teacher to simply a high-speed calculator performing data reduction; however, all forms of educational computer utilization can be broadly classified as Computer-Aided Instruction (CAIDI). This general classification can be further subdivided into Computer-Assisted Instruction (CAI), Computer-Evaluated Instruction (CEVIN), including the use of computer-generated repeatable exams, and Non-Interactive Computer Applications (NICA), using the criteria presented in Fig. 1. Further information on this classification scheme for educational computer usage is given in a previous review by this author [1]. This earlier review also covers methods for teaching programming languages and optimizing the role of NICA in the chemistry curriculum; consequently, these topics will not be discussed further. Instead, the emphasis in the current discussion will be on techniques and pedagogical strategies for "group" CAIDI; i.e., the effective instructional use of on-line computing in the classroom as a component of the lecture environment. This usage is neither CAI nor NICA, but rather incorporates some of the features of each. The extent to which on-line classroom computing becomes at least "pseudo-CAI" depends both on the extent to which a given program includes interrogational features and remedial branching, as well as on the extent to which students actively participate in the process. Without maximizing these characteristics, the principal goal of a classroom program execution is the expansion or clarification of course content; hence, the category is more nearly NICA. In either case, the machine is simultaneously interacting with an entire class, and the designation "group CAIDI" is given for on-line classroom computing.

II. HARDWARE REQUIREMENTS FOR ON-LINE CLASSROOM COMPUTING

To date, on-line computing during a lecture class has not been extensively used primarily because of a combination of economic, technological, and convenience factors. The process is, however, technically feasible, and a variety of possible configurations for presenting computer output in a visual form suitable for large student audiences is shown in Fig. 2. Methods which employ either a single CRT-equipped teletype as the viewing medium or several television monitors operated from a signal-generating CRT terminal are relatively expensive. Furthermore, the use of a single CRT teletype in the classroom presents viewing problems for audiences larger than 25 -30 students. Both methods also require frequent moving of the graphics terminal in and out of classrooms, which is neither convenient nor desirable. However, the recent development of a small, low-cost teletype add-on or adapter (Ann Arbor Terminals, Ann Arbor, Michigan) that provides for instantaneous video projection of the same output as displayed on the printed teletype copy has made on-line computing a more practical classroom technique. By having a room equipped with only a telephone line and suitably sized television set(s), an instructor can now conveniently supplement his lectures with the results of on-line computing using a portable teletype

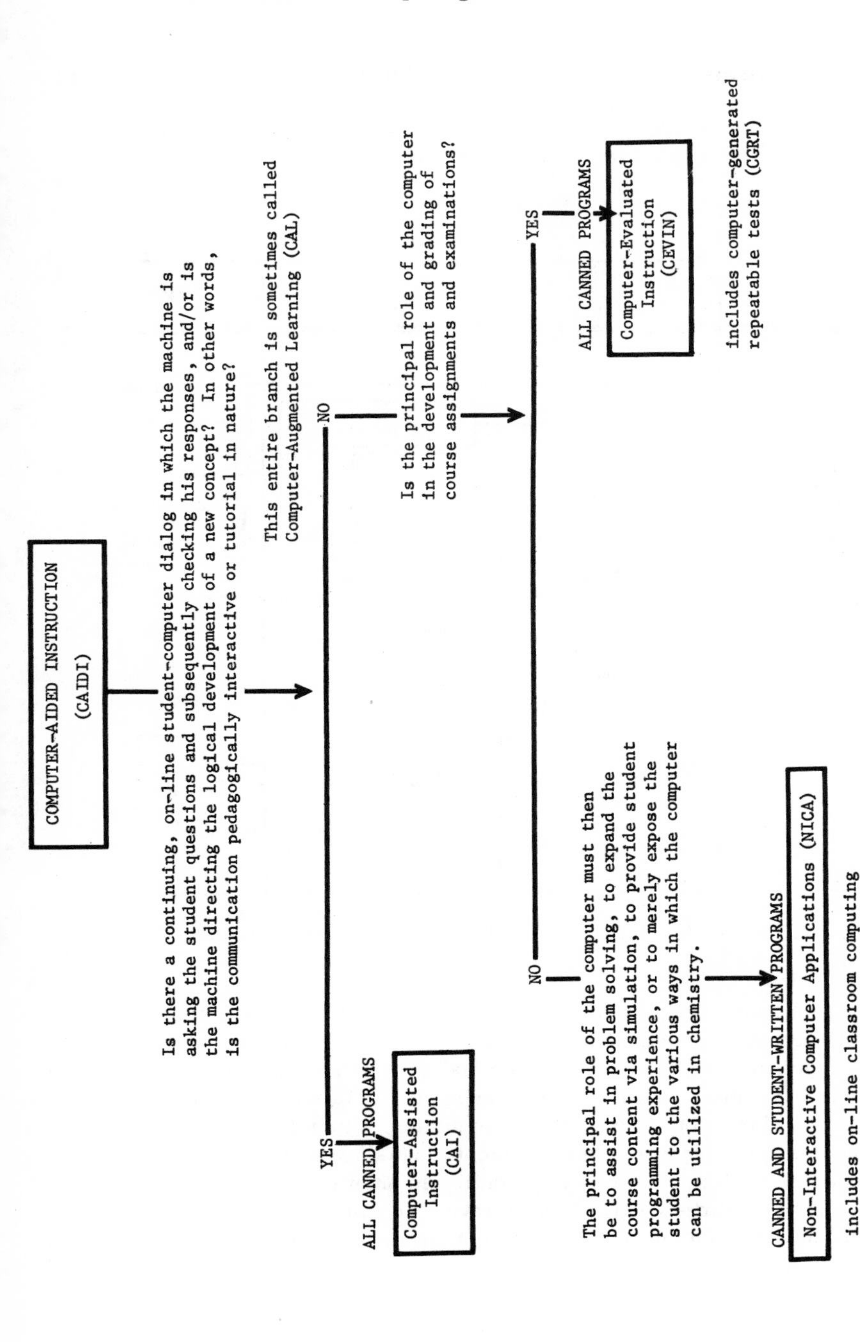

Fig. 1. Classifications for instructional computer usage.

terminal. The compactness of the equipment needed for on-line classroom computing using this method of video output display is shown in Fig. 3. Basically this device operates by generating a video signal from the same signal given to the teletype terminal by the computer.

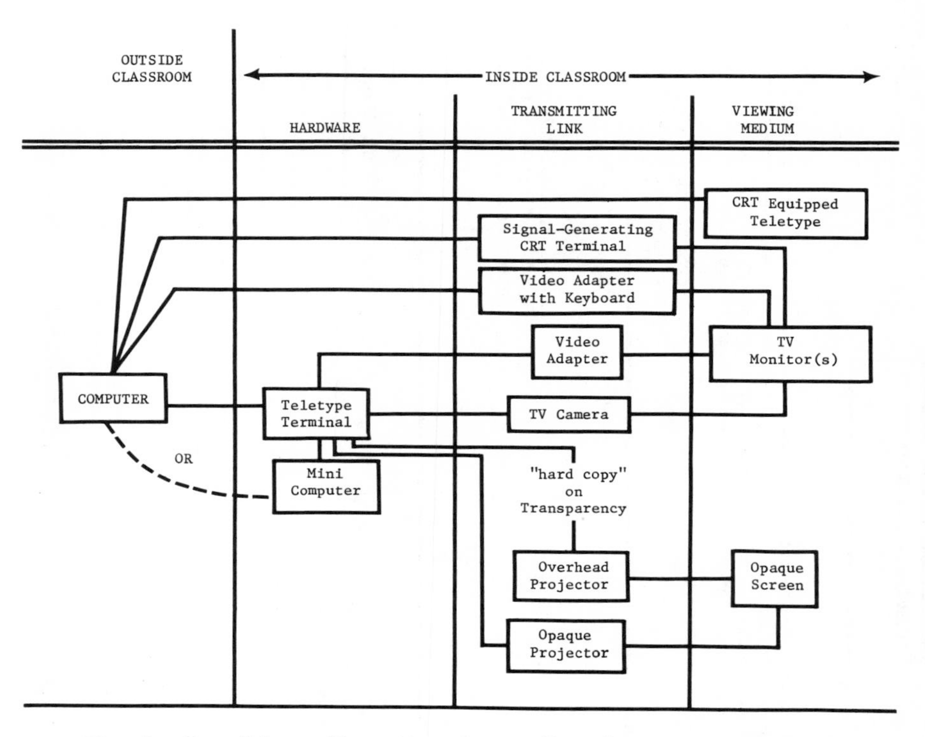

Fig. 2. Possible configurations for on-line classroom computing.

Television projection of computer output for group viewing can also be accomplished by focusing a TV camera on the printed teletype output. Although technically effective, this method normally requires that the instructor have an assistant present during his lecture and essentially converts the classroom into a television studio, which distracts somewhat from the desired educational environment. This arrangement for projecting the results of on-line classroom computing is shown in Fig. 4. Clearly, the television camera-teletype configuration occupies more space and is certainly less convenient to use in the classroom than the aforementioned video adapter.

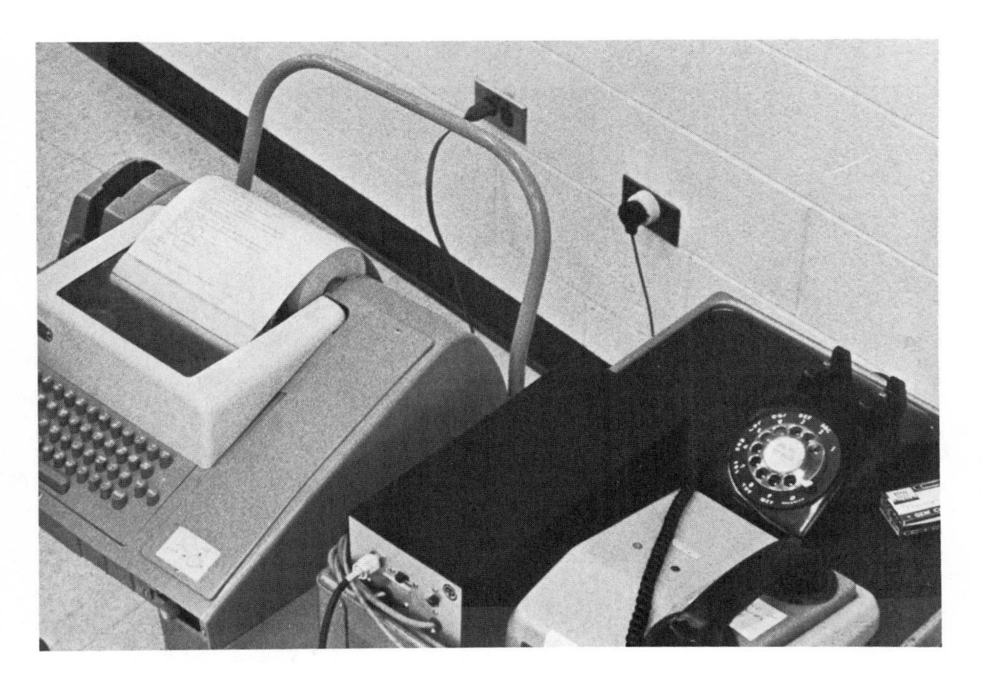

Fig. 3. Video adapter-teletype configuration for video projection.

In addition to these video-oriented techniques for group viewing of time-shared computer output, there are several methods based on optical projection of the teletype output onto an opaque screen. One method employs the teleprinter projector (I. P. Sharp Associates Limited, Ontario, Canada), which is essentially an overhead projector customized for use with a variety of teletypewriters. A roll of reusable Mylar is substituted for the paper in the terminal. Input/output is typed directly onto the Mylar which is advanced over a lens capable of projecting images onto walls, blackboards, or screens in any size up to 12 ft by 12 ft. Although this device permits one to write comments directly on the projected output, and to reverse the Mylar roll in order to review segments of the computed output, it appears that the video adapter offers more consistent viewing clarity with easier equipment preparation. For example, once a terminal is prepared for overhead projection it is necessary to remove the Mylar roll and replace it with a standard roll of paper prior to routine use of the teletype outside of the classroom. On the other hand, the add-on video adapter can merely be switched off without disconnecting, since it does not affect normal use of the teletype.

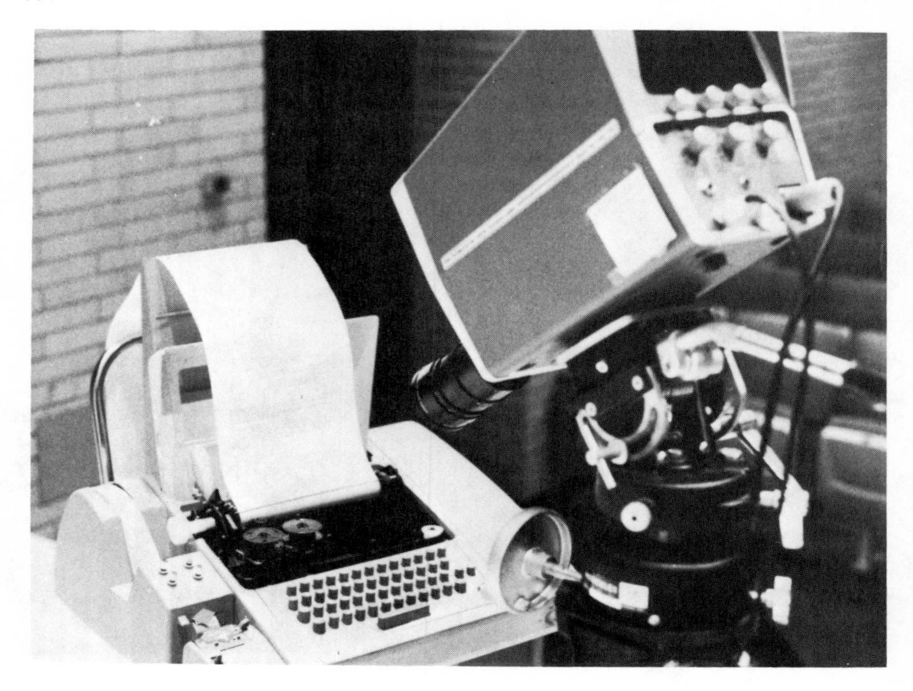

Fig. 4. TV camera-teletype configuration for video projection.

A second device for optical projection of teletype output is the commercially available TyProjector (Bolt Beranek and Newman, Inc., Santa Ana, California). This add-on unit, which mounts directly on a Model 33 Teletype with no mechanical modification, creates an image by reflecting light off the typewriter printout as it is being typed. This image may be displayed on any projection screen in front of the operator.

Both devices for optical projection are comparable in price to the more versatile video adapter, i.e., in the price range of $700-$1000. This adapter, measuring 5 × 5 × 15 in., is equipped with the appropriate cable sockets for easy input connection to a Model 33 Teletype and for output connection into a video receiver. This commercially available device is also equipped with a switch that permits the user to disable the teletype and merely get the video output without accompanying hard copy. The principal advantage of this capability is that the teletype is silent and the lecturer need not speak over the background of mechanical noise. The disadvantage, of course, is that there is no permanent record of the results. The adapter comes in several models, but the one currently being used for chemistry instruction at Eastern Michigan University is capable of displaying an image consisting of 16 vertical lines of output, each 32 characters in length. This has provided excellent viewing in a classroom with approximately 100 seats and four ceiling-mounted 25-in. television sets.

At present the adapter accommodates only alphanumeric output or "typewriter plots," and continuous functions cannot be graphically represented on video. Also, as with the teletype itself, the video display of chemical formulations is complicated by the absence of lower-case characters, subscripts, and superscripts. This is not a major problem, however, as students rapidly adjust to seeing chemical notations written in line form. The size of the individual characters on the 16 x 32 grid allows for easy viewing, and the rate of character display is identical to that of a teletype printing hard copy only. When the output exceeds the capacity of the viewing grid, the excess data can be accommodated by utilizing either the roll or page mode of the adapter. In roll mode the first line is eliminated, the second through the sixteenth lines move upward one space, and the new (or seventeenth) output line occupies the bottom position. This process is repeated as many times as necessary. In page mode the seventeenth line merely replaces the first, the eighteenth line replaces the second, and so on until all the output is displayed. The page mode is somewhat more confusing to students as the last lines of output are at the top of the TV display; however, either mode is satisfactory for classroom use. Furthermore, irrespective of which adapter mode is used, the spacing of the output displayed by a given program must be software-controlled, as the adapter does not provide for direct, real-time control.

III. DESIGN DESCRIPTION OF THE VIDEO ADAPTER

The function of the video adapter is to convert digital data, as received from a local or remote computer, into a video signal which will display that data on a conventional TV screen. Familiar applications outside the educational world include TV broadcast titling and airport flight arrival/departure displays. As an instructional device, the adapter interfaces the computer to a TV, providing a powerful classroom teaching aid.

A brief description of each design feature of the video adapter (Ann Arbor Terminals, Inc.) will be given, along with a complete detailed block diagram of one model.

A. Video Adapter Input

The video adapter plugs in in parallel with the teletypewriter. It prints on the TV screens whatever the teletypewriter prints. It receives data in standard teletypewriter form: bit-serial, start/stop synchronized, ASCII-coded (American Standard Code for Information Interchange; Fig. 5).

b	6 5 4	0 0 0	0 0 1	0 1 0	0 1 1	1 0 0	1 0 1	1 1 0	1 1 1
3210	↓ →	0	1	2	3	4	5	6	7
0000	0	NUL	DLE	SP	0	@	P		
0001	1	SOH	DC1	!	1	A	Q		
0010	2	STX	DC2	"	2	B	R		
0011	3	ETX	DC3	#	3	C	S		
0100	4	EOT	DC4	$	4	D	T		
0101	5	ENQ	NAK	%	5	E	U		
0110	6	ACK	SYN	&	6	F	V		
0111	7	BEL	ETB	'	7	G	W		
1000	8	BS	CAN	(	8	H	Z		
1001	9	HT	EM	)	9	I	Y		
1010	10	LF	SUB	*	:	J	Z		
1011	11	VT	ESC	+	:	K	[		
1100	12	FF	FS	,	<	L			
1101	13	CR	GS	–	=	M	]		
1110	14	SO	RS	.	>	N			
1111	15	SI	US	/	?	O	_		DEL

Fig. 5. Standard ASCII Code Set.

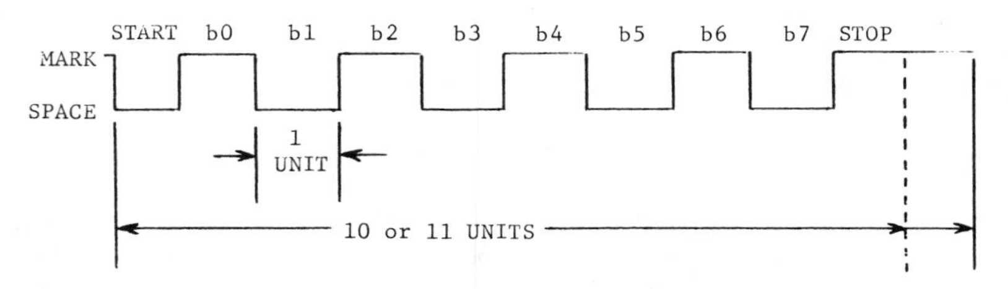

Fig. 6. Serial character transmission. (Waveform shown for letter "U" with even parity.)

Each character received is represented by an 11-bit code (Fig. 6). The teletypewriter line normally is at a digital "one" (MARK) level. When a character is transmitted, the line first goes to a "zero" (SPACE) level for one bit-time. This signals the video adapter (and the teletypewriter) that the next seven bits received (b 0-6) represent the character code as defined in Fig. 5. The next bit received (b7) is a "parity" bit, which may be used by the receiving device as an indicator of correct transmission. The line must then return to the "one" level for at least two bit-times before another character is transmitted.

B. Video Adapter Output

The video adapter plugs into the first classroom monitor in place of the normal closed-circuit TV input. (Note: It can also be plugged in in series with this input for superimposition of computer-generated data on the normal CCTV picture.) The computer-generated data automatically appear on all monitors in the classroom. Output signal strength is sufficient to drive up to a dozen monitors. Thus, it can accommodate even a very large auditorium.

The output signal is composite video in accordance with American Standard RS-170. This is compatible with all conventional U.S. TV monitors.

C. Video Adapter Operation

A simplified block diagram of the Ann Arbor Terminals' Model 204 video adapter is shown in Fig. 7. It contains four printed-circuit boards as described below. Its specifications are summarized in Fig. 8, and its code set is given in Fig. 9.

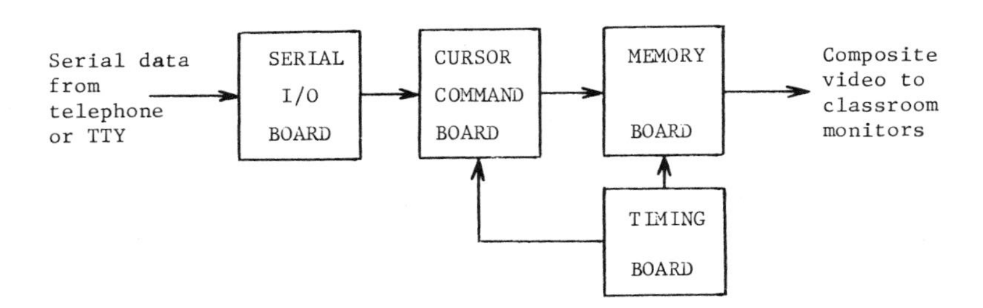

Fig. 7. Video adapter block diagram.

DISPLAY

DISPLAY DEVICE. Any conventional TV set or 525-line video monitor. Free-standing monitor optional.

DISPLAY FORMAT. 32 characters by 16 lines.

CHARACTER SET. 64 alphanumeric ASCII characters. See code table.

CHARACTER SHAPE. Defined by a 5x7 dot matrix in an 8x12 dot field.

CHARACTER SIZE. Proportional to screen size. Approx. 1/4" on 9" screen.

REFRESH RATE. 60 frames/sec.

CURSOR. Underline cursor.

OPERATING MODES

PAGE or ROLL mode, switch-selectable.

PAGE MODE OPERATION. Characters enter at cursor location. Cursor automatically advances on character entry. Advance past end of line advances to the beginning of next line. Advance past end of bottom line advances to beginning of top line. All commands active.

ROLL MODE OPERATION. Same as page mode, except that advance past end of bottom line moves all lines up one. All commands active.

PHYSICAL DATA

SIZE: 15 x 5 1/2 x 5 1/2 inches

WEIGHT: 8 lbs.

POWER: 115 volts, 60 Hz., 30 watts

COMMAND SET

CLEAR SCREEN (CLR). Moves the cursor to the first character position of the first line (upper left corner), and clears the screen. Key: Control L.

CURSOR HOME (HME). Moves the cursor to the first character position of the first line. The displayed data are unchanged. Key: Control K.

CARRIAGE RETURN (CR). Return cursor to the first character position of the line in which it is located. Key: RETURN.

CURSOR RIGHT (→). Moves cursor one character position to the right. If the last character position of a line, the cursor moves to the first character position of the next line. Key: Control I.

CURSOR LEFT (←). Moves cursor one character position to the left. If the first character position of a line, the cursor moves to the last character position of the line above. Key: Control H.

CURSOR UP (↑). Moves the cursor one line above its present position. If in the top line, the cursor moves to the same position in the bottom line. Key: Control N.

CURSOR DOWN (↓). Moves the cursor one line below its present position. If in the bottom line, the cursor moves to the same position in the top line. Key: LINE FEED.

CURSOR ADDRESS (CDR). Moves the cursor directly to the character position specified by the next two sequential data input bytes. The first byte specifies line address; the second byte specifies character address within the line. Line address is encoded as a hexadecimal digit in b3-b0 representing the 2's complement of line position 0-15, top to bottom on the screen. Character address is encoded as a 5-bit binary number in b4-b0 representing character position 0-31, left to right on the screen. The high-order bit positions of both bytes are "don't care".

Fig. 8. Specifications — Ann Arbor Terminals Model 204 video adapter.

The Timing Board generates the TV raster scan and refresh memory timing for 32 × 16 character format on the screen. (Note: Formats are available from 32 to 80 characters per line, and 8 to 24 lines.) The 32 × 16 character format is recommended for classroom use; it results in a character size compatible with normal closed-circuit TV group-viewing. The Timing Board also generates the timing for character-addressable memory load, generates cursor video, and provides for asynchronous input control of memory load.

b	6 5 4	0 0 0	0 0 1	0 1 0	0 1 1	1 0 0	1 0 1	1 1 0	1 1 1
3210	↓ →	0	1	2	3	4	5	6	7
0000	0			SP	0	@	P	@	P
0001	1			!	1	A	Q	A	Q
0010	2			"	2	B	R	B	R
0011	3			#	3	C	S	C	S
0100	4			$	4	D	T	D	T
0101	5			%	5	E	U	E	U
0110	6			&	6	F	V	F	V
0111	7			'	7	G	W	G	W
1000	8	←		(	8	H	X	H	X
1001	9	→		)	9	I	Y	I	Y
1010	10	↓		*	:	J	Z	J	Z
1011	11	HME		+	;	K	[	K	[
1100	12	CLR		,	<	L	\	L	\
1101	13	CR		-	=	M	]	M	]
1110	14	↑		.	>	N	^	N	^
1111	15	CDR		/	?	O	_	O	_

Fig. 9. Video adapter code set.

The Memory Board stores the full screen of alphanumeric data (512 characters for the 32 × 16 format) and generates the composite video output for display of the data on the classroom monitors.

The Cursor Board buffers the incoming data to memory, holds the cursor address, and mechanizes the roll/page and command set operations (Fig. 8).

The Serial Board converts the serial asynchronous input data from the teletypewriter line to bit parallel for the Cursor Board and Memory.

D. Detailed Design Description

1. Raster Scan Timing

The Timing Board creates a raster scan of 262 lines repeated 60 times per second. The horizontal and vertical sync frequencies are 15,720 Hz and 60 Hz, respectively.

The basic clock is a 6.288-MHz oscillator. The clock is divided by 8 in a bit counter (BC) to establish character time. The character is displayed at dot positions 1-5 of each character time; the remaining 3 dot positions form the inter-character spacing.

The character clock (BC2) is divided by 50 through a horizontal counter (HC) to establish the TV line time. This provides 50 character positions per TV line, 32 of which are used for display. The remaining character times are used for retrace and horizontal margins.

The TV line clock (LCP*) is decoded off the horizontal counter and divided by 12 in a line counter (LC) to establish the data line time. The characters are displayed on TV lines 1-7 of each data line; the remaining five TV lines form the interline spacing. The cursor, when used, is displayed on TV line 10.

The 4-TV-line clock (LCl) is divided through a vertical counter (VC) by 64. A 6-TV line dwell (DWL) is inserted at the end of the vertical counter to establish the 262-line frame at 60 Hz.

The raster allocation and timing are summarized in Fig. 10. The generation of horizontal sync at character count 38 and vertical sync at TV line 216 achieves the desired centering of the display.

2. Memory Timing

The refresh and buffer memories are shift registers recirculated in bit parallel in synchronism with the raster scan.

The buffer memory is 50 characters long, 32 of which are the data line being displayed. It is shifted continuously by the character clock, BC2, causing it to recirculate once each TV line.

The refresh memory is 512 characters long. It is shifted at character rate through 40 characters each TV line. During TV line 0 of each 12 TV lines, the contents of the refresh memory is copied into the buffer. The data transferred precesses by 32 characters, or one data line, each transfer.

The refresh memory recirculates once each 12.8 TV lines, or 20 times per frame, permitting it to be loaded asynchronously at up to 1200 characters per second.

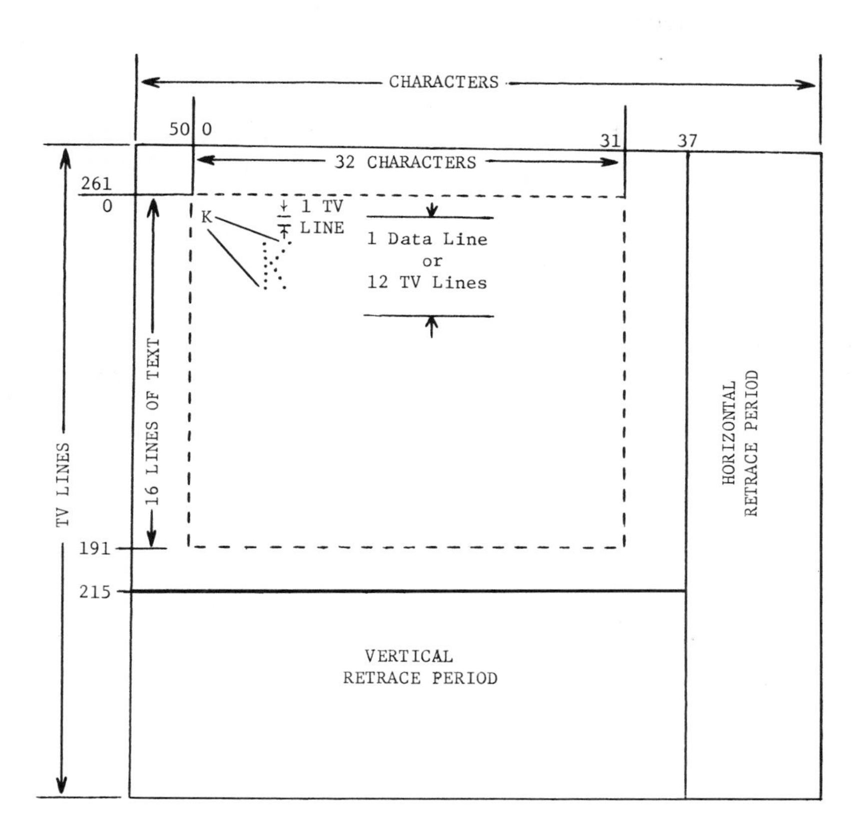

Fig. 10. Raster allocation.

3. Memory Address Timing

The refresh memory clock is counted in a memory address counter (MA). This is compared with a 9-bit address input (AR) to produce two pulse trains time-phased with the recirculation of the memory. The load clock (LDC*) train marks the time at which the addressed contents of memory can be written or read. The cursor video (CRV) train marks the time at which the addressed character is underlined on the screen.

The character address input, AR 0-4, is encoded as a 5-bit binary number representing character position 0-31, left to right on the screen. The line address input, AR5-8, is encoded as a hex digit representing the 2's complement of line position 0-15, top to bottom on the screen.

4. Load Control

The load control logic provides once-only gating for the load clock. Each time the load circuit is strobed (STB*) the next, and only the next, load clock is gated out as a load pulse (LDO*). A load pulse delayed one character time (LDD) is also generated. A "ready" output (RDY) goes FALSE at strobe and remains FALSE until completion of the load pulse.

Two dc control inputs are provided. The load inhibit (LDI*) input prevents the load circuit from being strobed. The repeat (RPT*) input causes the circuit to automatically strobe at 1200 characters/sec.

5. Memory Board

The Memory Board stores the full screen (512 characters) of alphanumeric data and generates the composite video output for display of the data on the classroom monitors.

6. Refresh Memory

The refresh memory contains six 512-bit MOS dynamic shift registers recirculated in bit parallel in synchronism with the raster scan. Data are loaded (DI0-5) and read (DO0-5) as 6-bit parallel bytes.

Load is accomplished by pulsing the memory load input (MLD*) FALSE in sync with the load clock (LDC*) generated on the timing board. This breaks the memory recirculation loop and causes the data at the memory input to be written into memory.

The memory recirculates 20 times per frame (1/60 sec) permitting 1200 random accesses per second.

7. Buffer Memory

The buffer memory contains six 50-bit MOS dynamic shift registers recirculated in bit parallel once each TV line. The memory holds the 32 characters being displayed, plus 18 characters of "garbage" hidden by the horizontal retrace blanking.

Every 12 TV lines, a new set of characters is copied from the refresh memory into the buffer. The relative precession between the memories is such that the data copied are the next data line to be displayed.

8. Video Generator

The video generator contains a 2560-bit MOS read-only-memory (ROM) and a 5-bit video output shift register. The ROM stores the 5 × 7 dot matrix representation of each of the 64 displayable characters. Organization is 9-bit address in, 5-bit row video out. The address is composed of the 6-bit character code, selecting which character is to be displayed, and the 3-bit TV line count (LC), selecting which row of the matrix is to be read out. The 5-bit readout is strobed into the video output register by the row video strobe (RVS) signal from the Timing Board and shifted out serially by the dot clock (CP*) as video.

9. Video Amplifier and Modulator

The video amplifier adds together two sources of video and the horizontal and vertical sync pulses to form the composite video output (CVD) required by the TV monitors. One source of video is the alphanumeric video from the video generator; the other source is external (EVD), usually tied to the cursor video output of the Timing Board.

10. Cursor Board

The Cursor Board (1) buffers incoming data to memory, (2) holds cursor address, and (3) mechanizes the roll/page and command set operations (Fig. 6).

11. Data Register

A 7-bit-parallel data register buffers incoming data to the refresh memory. A character is loaded into the register on the trailing edge of strobe (STB*). It is held until loaded into memory if a "display" character, or until the command is effected if a "command" character. Logic between the data register and the memory modifies data bit-5 so that incoming lower-case characters are stored and displayed as upper-case characters.

12. Address Register

A 9-bit address register (AR), connected to the address input of the Timing Board, determines the location in memory at which the characters are loaded.

13. Command Control Logic

The command control logic operates on the address register to control the load location. All actions are done in synchronism with the load pulse (LDO*) input. If the character in the register is a "display" character (DI5+DI6) LDO* is gated out as a memory load pulse (MLD*) to load the data into memory, and gated through the command structure to advance the address register. If the data in the register is a "command" character (DI5*, DI6*), LDO* is gated through a command decoder to effect the commanded action. Cursor Home clears the address register. Carriage

Return clears the low-order 5-bits of the register. Cursor Right increments the register, Cursor Left decrements it. Cursor Up increments the high-order 4-bits, Line Feed decrements it. Clear Screen clears the address register, loads the data register with space code (SPC*), and holds the MLD* output FALSE so that spaces are loaded into the refresh memory, clearing it until the next character is entered.

The Cursor Address operates through a state counter which selectively gates the first byte following the command (line address) into the high-order 4-bits of the address register and the second byte (character address) into the low-order 5 bits of the register.

The command logic operates in page or roll mode depending on the state of the RMD* input: FALSE = Roll mode, TRUE = Page mode. In roll mode, the line counter is inhibited from advancing past the bottom line. Instead, a roll state counter is actuated which first clears (fills with spaces) the top line of the screen and then inhibits the refresh memory clock (MCI*) for 32 character times, which rotates the screen up one line in respect to picture sync.

14. Serial I/O Board

The Serial I/O Board converts serial asynchronous data to parallel on input, and parallel data to serial asynchronous on output.

15. Serial Data Input

The board accepts bit-serial, character-asynchronous input data in either 10- or 11-unit code. Waveform is drawn (Fig. 6) for the letter "U" with even parity. Parity (b7) is ignored on input.

The data bits are clocked into a serial-to-parallel register by a start/stop receive oscillator. The oscillator is started by the incoming START pulse, shifts the data into the register at its preset band rate, and stops until the next START pulse.

16. Serial Data Output (keyboard models only)

Data are transmitted as an 11-unit code, with parity transmitted at MARK. Bit-parallel data at the keyboard input (KI) are loaded into the parallel-to-serial register at the leading edge of strobe (STB*). They are clocked out serially by the free-running transmit oscillator.

IV. PEDAGOGICAL ASPECTS OF ON-LINE CLASSROOM COMPUTING

The direct real-time use of computing within the framework of a classroom lecture has many advantages. Included among the obvious benefits of this instructional technique are the following:

1. It provides for rapid, accurate calculations and neat, orderly display of the results.

2. It permits the use of sophisticated, accurate mathematical methods rather than the approximations often used for convenience.

3. It introduces the students to computer programs which can subsequently be used for outside assignments.

4. It provides a focus for active classroom discussion, i.e., via a multimedia three-way instructor/student/computer interaction.

5. It brings a dynamic, modern "instrument" into the classroom.

The provision for rapid, accurate calculations along with the possible application of sophisticated mathematical methods — a direct result of the rapid numerical processing capability of the computer — serves to upgrade the mathematical foundation on which a particular course is built, often to levels far beyond that normally associated with the course number itself. For example, freshman courses in chemistry seldom require more mathematics than basic algebra along with perhaps a few concepts from calculus. Through the use of on-line classroom computing, students in freshman general chemistry can be exposed to an applications-oriented use of iteration techniques along with other numerical methods. Although it is true that an in-depth knowledge of the mathematics involved is not conveyed to the students, it is certainly also true that this does not inhibit their ability to see the obvious benefits derived from the application of such methods to chemical problems. This strategy is not a pedagogical compromise; it merely emphasizes the role of mathematics as an applied tool for chemists. The convenient implementation of this tool via the computer is simply making full use of available technology.

Not only does the vehicle of computing serve to bring sophisticated mathematics into elementary chemistry classrooms, but it also serves the dual purpose of exposing an entire class to the concept of applied computer programming. Programs introduced to students in the classroom can subsequently be used as the basis for rigorous homework assignments or "take-home" exams. In this instance, it is assumed that the in-class usage of a given program has served both to illustrate its proper application to a given type of chemical problem and to familiarize the student with the input requirements of the program.

In addition, on-line classroom computing with video output display constitutes an excellent focus for active class discussion. This unrehearsed, unstructured approach to real-time problem-solving enables the student to become maximally involved in the instructional process. Students need not just passively observe the instructor solving a problem of his prior choosing, but rather, in this situation, the input parameters can be student-chosen.

The instructor need only provide sufficient direction to insure within a reasonable number of program executions that the desired principle or correlation is indeed being illustrated. Dialogue is prompted not only between individual students and the instructor, but also among students, as disagreements sometimes arise concerning the best set of input parameters to illustrate a given point. This active student participation seems to promote a desire in many students to be better prepared for class so that they can be an integral part of the computing activity.

Finally, just the added dimension of having a modern, relevant device or "instrument" in the classroom seems to enliven the lecturing environment. Of course, as is always true when dealing with a form of instructional technology, it is extremely important that the machine or device involved does not provide simply a theatrical effect. To some extent such an effect is tolerable, though; anything which adds excitement or creates interest invariably stimulates greater student attention which in turn leads to enhanced learning. Nevertheless, in order to maximize the pedagogical value of on-line classroom computing, certain rules should be observed. Based on experience to date, this technique is most effective provided that:

1. It is not used too frequently; i.e., if on-line computing is used during every lecture in a given course, then just as with any other instructional aid, students will become somewhat bored with its use.

2. The applications are carefully selected so as to truly reinforce or expand upon some topic under discussion. It is imperative that the data reduction or simulations performed be those which truly require mathematical accuracy not otherwise obtainable by approximation methods, or that they require the rapid use of mathematical operations (e.g,, iteration techniques) not otherwise possible in real time during a lecture. In other words, the computing medium is not the complete message in itself, and the technique must not be exploited purely for its theatrical value during a lecture.

3. A rapid response time-sharing network is employed which provides access to a reasonably large library of stored computer programs.

4. The students have access to the same programs for use outside of the classroom. This can be accomplished either by using the identical program in time-shared mode or by providing an analogous batch version of the program. In either event it is very important that any interested student be able to experiment on his own with the same type of calculation performed during the lecture. This not only permits repetition if necessary but it also allows the student to pursue other examples of his own choosing.

The hard copy from the classroom usage should be posted or copied and distributed to all students for their reference. Also, programs for on-line classroom use should use minimum input, so that the students do not become bored during prolonged typing sessions by the instructor. It is likewise important that the instructor type with reasonable accuracy, since all errors even though removed from the actual data entry by appropriate software controls, remain displayed on the video screen. In other words, although typing errors are unavoidable and can be easily deleted from the machine record, they are somewhat distracting when displayed on video in this computing mode.

V. PROGRAMS USED FOR ON-LINE CLASSROOM COMPUTING IN GENERAL CHEMISTRY

The potential applications of this technique for chemistry education are virtually unlimited and certainly too numerous to list. In fact, considering the proposed subdivision of NICA into five categories (Fig. 11), it is relatively easy to suggest applications and the accompanying courses to fit each category. For example, the real-time, on-line demonstration of data evaluation via comparison to computer-stored databanks as well as the demonstration of computerized information retrieval via keyword interest profiles would be excellent supplements to conventional lecture discussions of those topics. Nevertheless, the principal applications of on-line classroom computing would seem to come in the areas of data reduction and simulation. Furthermore, if the goal underlying the use of this form of educational technology is to improve chemistry instruction in those course areas where the greatest number of students are involved, then attention should be focused on the general chemistry course along with other lower level offerings.

To illustrate the type of computer programs and strategies which have been found to be effective for on-line classroom computing in general chemistry, several examples will be discussed.

1. PHHAC

This program is designed to calculate the pH of acetic acid solutions both with the aid of simplifying assumptions and via the solution of a quadratic equation. As can be seen from the listing of the program (see Fig. 12), the algorithm is a simple one and the programming (FORTRAN for a GE 400 used by Applied Computer Time Share, Detroit, Michigan) is elementary. Nevertheless, this program has been very successful in illustrating several basic points in the solving of problems dealing with weak acids and ionic equilibrium. For example, as shown in the sample output (Fig. 13), the effect of initial molar acetic acid concentration on the validity of the simplifying assumption is easily demonstrated. This can be illustrated over a narrow concentration range (e.g., 1×10^{-5} to 1×10^{-6} M) or over a much wider range (e.g., 1×10^{-1} to 1×10^{-5}) as shown in Fig. 13 by the use of the decrementing or multiplying option which is available. Also, in a deliberate attempt to provoke student thought and dialogue, the program was designed to calculate erroneous pH values for extremely dilute acid solutions. As illustrated in Fig. 13, basic pH values are calculated when acid concentrations of 1×10^{-7} M or less are specified. This result was used both to stimulate class discussion on the mathematical handling of such acid solutions, and to form the basis of an outside assignment on the subject.

Category	Description
1. DATA REDUCTION	Mathematical processing of data, usually based on a laboratory experiment performed by the student.
2. DATA SIMULATION	Generation of tabular or graphical data based on theory rather than the student's own experiment; e.g., calculation and plotting of electron densities in atoms and molecules based on quantum mechanical equations.
3. DATA EVALUATION	Comparison of experimentally-obtained data to a library of reference data stored in the computer; e.g. identification of a compound via the comparison of its x-ray diffraction powder patterns to the complete ASTM library of reference powder patterns.
4. DATA ACCUMULATION AND OPTIMIZATION	Interfacing of a computer with an experimental apparatus; i.e., using the computer to control the actual performance of an experiment and to continuously collect and, if desired, monitor the quality of the data. Examples include on-line computer-automated x-ray diffraction equipment and mass spectrometers.
5. INFORMATION RETRIEVAL	Indexing and searching the modern research literature with the aid of computerized information retrieval techniques based on the use of descriptive keywords or chemical formulations.

Fig. 11. Five categories of pedagogically noninteractive computer applications (NICA).

In addition to the specific features of program PHHAC, this program should also serve to direct attention to several general features of all programs used in on-line classroom mode. For example, all such programs are structured with an interrogational beginning, thereby leading the user via a series of questions to the arithmetic execution phase, whether or not the individual has any programming knowledge. The initial question "How

PHHAC

```
10 PRINT "FOR TV TYPE TV"
20 PRINT "FOR TELETYPE TYPE TTY";
30 INPUT A$
33 L1=0
34 Z9=0
35 W=0
40 I=1
50 IF A$="TV" THEN 100
51 PRINT "DO YOU WANT A DESCRIPTION OF THIS PROGRAM, TYPE YES OR NO";
52 INPUT Z$
53 IF Z$="YES" THEN 870
60 GO TO 110
100 I=0
110 PRINT "NAME OF ACID";
120 INPUT B$
130 PRINT"KA OF ";B$;" IS ";
140 INPUT K
150 B=K**2
155 DEF FNL(X)=(-LOG(X))/2.3026
160 PRINT "INITIAL CONCENTRATION";
170 INPUT C
180 IF I=1 THEN 220
190 PRINT "MULTIPLYING FACTOR";
200 GO TO 230
220 PRINT "MULTPLYING FACTOR, EACH CONCENTRTION WILL"
221 PRINT "BE MULTIPLIED BY THIS FACTOR.";
230 INPUT D
240 PRINT "FINAL CONCENTRATION";
250 INPUT F
260 IF F>C THEN 280
270 GO TO 291
280 PRINT "FINAL CONCENTRATION MUST BE"
281 PRINT "SMALLER THAN INITIAL"
290 GO TO 240
291 IF I=1 THEN 295
292 PRINT "  CONC.     PHA     PHB     PHC"
294 GO TO 298
295 PRINT"CONCENTRATION OF ACID    PHA       PHB       PHC       PHD"
298 FOR M=1 TO 30
299 IF C<F THEN 810
300 IF C<=1.0E-11 THEN 810
305 H1=SQR(K*C)
310 P1=FNL(H1)
320 H2=(-K+SQR(B+(4.0*K*C)))/2
321 IF H2<=0 THEN 340
330 P2=FNL(H2)
331 GO TO 350
333 GO TO 350
340 H2=1.0E-07
341 P2=99.999
342 Z9=1
350 H=H2
360 F1=(H**3)+K*(H**2)-H*(K*C+1.0E-14)-K*1.0E-14
380 F2=3*(H**2)+2*K*H-(K*C+1.0E-14)
390 H3=H-(F1/F2)
```

Fig. 12. FORTRAN listing of program PHHAC.

```
391 IF H3<=0 THEN 471
400 P6=FNL(H)
410 P7=FNL(H3)
420 D1=P6-P7
430 D2=ABS(D1)
440 IF D2<0.001 THEN 480
460 H=H3
470 GO TO 360
471 P3=7.00
472 GO TO 490
480 P3=FNL(H3)
490 IF H3<=0 THEN 493
491 H=H3
492 GO TO 500
493 H=1.0E-07
500 A=C/((H/K)+1)
510 H4=A+1.0E-14/H
520 P8=FNL(H)
530 P9=FNL(H4)
540 D3=P8-P9
550 D4=ABS(D3)
560 IF D4<0.001 THEN 600
570 H=H4
580 GO TO 500
600 P4=FNL(H4)
620 IF I=0 THEN 700
630 IF P1>=7.00 THEN 752
640 IFP2>=7.00 THEN 752
650 PRINT USING 680,C,P1,P2,P3,P4
680 :        ##.##↑↑↑↑    ###.###    ###.###    ###.###    ###.###
685 GO TO 760
700 PRINT USING 750,C,P1,P2,P3
710 L1=L1+1
720 IF L1=6 THEN 730
725 GO TO 760
730 PRINT "TO CONTINUE TYPE C";
735 INPUT G$
740 L1=0
750 :##.##↑↑↑↑ ##.### ##.### ##.###
751 GO TO 760
752 W=1
754 PRINT USING 755,C,P1,P2,P3,P4
755 :        ##.##↑↑↑↑    ###.###*  ###.###*  ###.###    ###.###
760 C=C*D
800 NEXT M
810 IF Z9=1 THEN 940
811 IF W=1 THEN 850
820 GO TO 900
850 PRINT
851 PRINT "* HOW CAN AN ACIDIC SOLUTION HAVE A PH LARGER"
852 PRINT "THAN 7.00?  TO ACCOUNT FOR THE IONIZATION OF WATER"
853 PRINT " STUDY PAGES 252-263 IN IONIC REACTIONS AND EQUILIBRIA"
```

Fig. 12. — Continued.

```
854 PRINT "BY OMER ROBBINS JR."
860 PRINT
869 GO TO 900
870 PRINT "THIS PROGRAM CALCULATES THE PH OF A WEAK ACID"
871 PRINT "BY FOUR DIFFERENT METHODS. CA IS THE INITIAL"
872 PRINT "CONCENTRATION OF ACID. THE METHODS ARE:"
873 PRINT "METHOD A - PHA - H+ CONC. IS THE SQUARE ROOT"
874 PRINT "OF (KA)(CA)."
875 PRINT "METHOD B - PHB - H+ CONC. IS FOUND BY SOLVING THE"
876 PRINT "QUADRATIC     (H+)**2 + (KA)(H+) - (KA)(CA) = 0."
877 PRINT "METHOD C - PHC - THE IONIZATION OF WATER IS ACCOUNTED"
878 PRINT "FOR AND THE H+ CONC. IS FOUND BY SOLVING A THIRD ORDER"
879 PRINT "POLYNOMIAL BY THE NEWTON-RAPHSON METHOD."
880 PRINT "METHOD D - PHD - SIMILAR TO METHOD C, BUT AN ITERATION"
881 PRINT "PROCEDURE IS USED TO SOLVE FOR THE H+ CONC."
899 GO TO 60
900 PRINT "IF YOU WANT TO CONTINUE"
910 PRINT "TYPE YES";
920 INPUT D$
930 IF D$="YES" THEN 10
935 GO TO 1000
940 PRINT "A PH OF 99.999 INDICATES AN ATTEMPT TO TAKE THE LOG"
941 PRINT "OF A NEGATIVE NUMBER."
960 GO TO 811
1000 END
```

Fig. 12. — Continued.

many times do you want to use this program?" is included merely to permit multiple executions without repeating the introductory comments, if any. The comment "To proceed — Type 1" is included to hold the output in viewing range on the video monitor(s). This was necessary in order to override a systems software feature which provided for the skipping of six to eight lines following completion of the execution.

Program PHHAC also illustrates the "pseudo-CAI" nature of properly designed software for on-line classroom use. Although the program does not provide for a check on the validity of student input data as would a true CAI program, it does emphasize the logical sequence of steps and the exact information required to solve the problem. Consequently, when used outside the classroom this program cannot provide tutorial help to students complete with grading or evaluation, but it can provide valuable help with the strategy of problem solving on an individualized basis. The only mistake which a student can make when using PHHAC or a similar program is to enter an erroneous numerical value as input.

```
PHHAC


FOR TV TYPE TV
FOR TELETYPE TYPE TTY ?TV
NAME OF ACID ?ACETIC ACID
KA OF ACETIC ACID IS  ?1.8E-05
INITIAL CONCENTRATION ?.1
MULTIPLYING FACTOR ?.1
FINAL CONCENTRATION ?1.0E-12
  CONC.      PHA     PHB     PHC
 1.00E-01   2.872   2.875   2.875
 1.00E-02   3.372   3.382   3.382
 1.00E-03   3.872   3.901   3.901
 1.00E-04   4.372   4.464   4.464
 1.00E-05   4.872   5.145   5.145
 1.00E-06   5.372   6.022   6.018
TO CONTINUE TYPE C ?C
 1.00E-07   5.872   7.002   6.793
 1.00E-08   6.372   8.000   6.978
 1.00E-09   6.872   9.000   6.998
 1.00E-10   7.372  10.000   7.000
```

```
FOR TV TYPE TV
FOR TELETYPE TYPE TTY ?TTY
DO YOU WANT A DESCRIPTION OF THIS PROGRAM, TYPE YES OR NO ?YES
THIS PROGRAM CALCULATES THE PH OF A WEAK ACID
BY FOUR DIFFERENT METHODS. CA IS THE INITIAL
CONCENTRATION OF ACID. THE METHODS ARE:
METHOD A - PHA - H+ CONC. IS THE SQUARE ROOT
OF (KA)(CA).
METHOD B - PHB - H+ CONC. IS FOUND BY SOLVING THE
QUADRATIC    (H+)**2 + (KA)(H+) - (KA)(CA) = 0.
METHOD C - PHC - THE IONIZATION OF WATER IS ACCOUNTED
FOR AND THE H+ CONC. IS FOUND BY SOLVING A THIRD ORDER
POLYNOMIAL BY THE NEWTON-RAPHSON METHOD.
METHOD D - PHD - SIMILAR TO METHOD C, BUT AN ITERATION
PROCEDURE IS USED TO SOLVE FOR THE H+ CONC.
NAME OF ACID ?ACEI\I\TIC ACID\D\D
KA OF ACETIC ACID IS  ?1.8E-05
INITIAL CONCENTRATION ?.1
MULTPLYING FACTOR, EACH CONCENTRTION WILL
BE MULTIPLIED BY THIS FACTOR. ?.1
FINAL CONCENTRATION ?1.0E-12
CONCENTRATION OF ACID     PHA        PHB        PHC        PHD
          1.00E-01       2.872      2.875      2.875      2.875
          1.00E-02       3.372      3.382      3.382      3.382
          1.00E-03       3.872      3.901      3.901      3.901
          1.00E-04       4.372      4.464      4.464      4.464
          1.00E-05       4.872      5.145      5.145      5.145
          1.00E-06       5.372      6.022      6.018      6.018
          1.00E-07       5.872*     7.002*     6.793      6.793
          1.00E-08       6.372*     8.000*     6.978      6.978
          1.00E-09       6.872*     9.000*     6.998      6.998
          1.00E-10       7.372*    10.000*     7.000      7.000

* HOW CAN AN ACIDIC SOLUTION HAVE A PH LARGER
THAN 7.00?  TO ACCOUNT FOR THE IONIZATION OF WATER
 STUDY PAGES 252-263 IN IONIC REACTIONS AND EQUILIBRIA
BY OMER ROBBINS JR.
```

Fig. 13. Sample output from program PHHAC.

2. KSP

This program permits the calculation of the solubility product constant of a salt if the solubility in grams per 100 ml of water is known, or alternately, both the gram and molar solubility if the Ksp value is known. As with the previously discussed PHHAC, this program is constructed upon a rather simple algorithm (see listing in Fig. 14), and it contains the same design features necessary for effective on-line classroom usage. The question-and-answer format not only leads the user to the computation itself, but it focuses student attention on the data required to perform the calculation.

```
KSP


10 PRINT "THIS PROGRAM WILL CALCULATE"
20 PRINT "KSP OR SOLUBILITY OF A"
30 PRINT "COMPOUND OF THE TYPE 'MA'"
40 PRINT "WHERE:"
50 PRINT "        M IS THE METALLIC CATION "
60 PRINT "AND     A IS THE ANION."
100 PRINT "HOW MANY TIMES WILL YOU USE"
110 PRINT "THIS PROGRAM ";
120 INPUT N
130 FOR I=1 TO N
140 PRINT "WILL CALCULATION BE KSP"
150 PRINT "OR SOLUBILITY "
160 PRINT "TYPE KSP OR SOL ";
170 INPUT M$
180 PRINT "WHAT IS FORMULA OF THE"
190 PRINT "COMPOUND ";
200 INPUT C$
210 PRINT "WHAT IS THE MOLECULAR WEIGHT"
220 PRINT "OF THE COMPOUND ";
230 INPUT W
240 PRINT "HOW MANY CATIONS ";
250 INPUT C
260 PRINT "HOW MANY ANIONS ";
270 INPUT A
280 IF M$="SOL" THEN 410
290 PRINT "WHAT IS THE SOLUBILITY IN"
300 PRINT "IN GMS/100 ML. H20 ";
310 INPUT S
320 LET X=10.0*S/W
330 LET S1=C**C*A**A*((X**(C+A)))
340 PRINT "FORMULA IS ";C$
350 PRINT "MOLAR SOL. IS ";X;" MOLES/L."
355 PRINT "KSP IS ";S1
360 GO TO500
410 PRINT "WHAT IS THE KSP ";
420 INPUT S3
430LET X=(S3/(C**C*A**A))**(1.0/(C+A))
440 LET S4=X*W/10.0
450 PRINT "FORMALA IS ";C$;" KSP IS ";S3
460 PRINT "SOL. IS ";S4;" GMS./100 MLS."
500 NEXT I
600 END
```

Fig. 14. FORTRAN listing of program KSP.

```
KSP


THIS PROGRAM WILL CALCULATE
KSP OR SOLUBILITY OF A
COMPOUND OF THE TYPE 'MA'
WHERE:
        M IS THE METALLIC CATION
AND     A IS THE ANION.
HOW MANY TIMES WILL YOU USE
THIS PROGRAM  ?2
WILL CALCULATION BE KSP
OR SOLUBILITY
TYPE KSP OR SOL  ?KSP
WHAT IS FORMULA OF THE
COMPOUND  ?AS2S3
WHAT IS THE MOLECULAR WEIGHT
OF THE COMPOUND  ?246.02
HOW MANY CATIONS  ?2
HOW MANY ANIONS  ?3
WHAT IS THE SOLUBILITY IN
IN GMS/100 ML. H2O  ?.00005
FORMULA IS AS2S3
MOLAR SOL. IS  2.03236E-6  MOLES/L.
KSP IS  3.74474E-27
WILL CALCULATION BE KSP
OR SOLUBILITY
TYPE KSP OR SOL  ?SOL
WHAT IS FORMULA OF THE
COMPOUND  ?CAF2
WHAT IS THE MOLECULAR WEIGHT
OF THE COMPOUND  ?78.08
HOW MANY CATIONS  ?1
HOW MANY ANIONS  ?2
WHAT IS THE KSP  ?4.0E-11
FORMALA IS CAF2 KSP IS  4.00000E-11
SOL. IS  1.68218E-3  GMS./100 MLS.
```

Fig. 14. — Continued.

The sample output from program KSP shown in Fig. 14b reveals also the advantage which the program offers in rapidly solving the higher degree exponential equations encountered in solubility product problems.

3. EO

This program (listing seen in Fig. 15) processes half-cell reactions and their corresponding electrode potentials to give the thermodynamically favored overall cell reaction. The output includes (see Fig. 16) the balanced equation and the cell EMF, and the correct result is obtained irrespective of the order in which the two half-cells are input to the program. This program was very effective in illustrating to students the various conditions under which a given chemical species could reverse its oxidizing-reducing function. Special attention should be directed to the comment statement at the beginning of program EO which acknowledges the efforts of a student who worked on modifying the program. This comment is permitted for any student making a significant modification or improvement in an existing program. The use of this public credit seemed to be an effective reward system. This discussion also leads into the subject of the use of on-line classroom computing as a technique for either teaching programming or at least stimulating interest in the methodology of program writing. Although no specific attempt has been made to teach programming per se via this method, instructors using these programs have commented on the general features of the programs, such as looping and testing statements. This has served to catalyze the desire in some students to enroll in a programming course or to learn the fundamentals of FORTRAN by self-study.

```
      TYPE 6001
6001 FORMAT (' IN THIS PROGRAM , FOLLOWING OXIDATION POTENTIAL'/
    +' CONVENTION IS USED:',/
    1' RED.FORM = OXD.FORM + (N)E-   EO')
      TYPE 6003
6003 FORMAT ('   A = B + (N1)E-     E1'/
    1 '   C = D + (N2)E-     E2'//)
      TYPE 6004
6004 FORMAT (' THIS PROGRAM AS DESIGNED WILL NOT TAKE'/
    1 ' CARE OF THE HALF REACTION'//
    2 ' H2 = 2H+ + 2E- EO = 0.0'//
    3  ' DO NOT USE THE ABOVE MENTIONED'/
    4 ' REACTION AS ONE OF YOUR INPUT.'//)
      TYPE 6005
6005 FORMAT ( ' APPROXIMATELY HOW MANY TIMES'/
    1 ' WILL YOU USE THIS PROGRAM ',$)
      ACCEPT 6000,J
      DO 5001 I=1,J
      TYPE 6012
6012 FORMAT ( ' WHAT ARE YOUR A, B, N1 AND E1?',/)
      ACCEPT 6020, A,B,N1,E1
6020 FORMAT (2A4,2G)
      TYPE 6030
6030 FORMAT (' WHAT ARE YOUR C, D, N2 AND E2?',/)
      ACCEPT 6020,C,D,N2,E2
      X=ABS(E1-E2)
      IF (N1-N2) 460,2000,460
 460 IF (E1-E2)162,260,362
 162 TYPE 6040
6040 FORMAT ( ' IS ONE OF THE SPECIES DIATOMIC? YES(1),NO(0). ',$)
      ACCEPT 6000,M
      IF (M) 160,160,363
 363 N=N1*N2
      TYPE 1001,N1,C,N2,B,N,A,N1,D,X
      GO TO 999
 160 TYPE 1001,N1,C,N2,B,N2,A,N1,D,X
1001 FORMAT (//I2,A4,' + 'I2,A4,' = 'I2,A4,' + 'I2,A4,3X,'CELL
    +EMF = ' F6.3,' VOLTS ')
      GO TO 999
```

Fig. 15. FORTRAN listing of program EO.

```
 362 TYPE 6040
     ACCEPT 6000,M
     IF (M)360,360,463
 463 N=N1*N2
     TYPE 1001,N2,A,N1,D,N2,B,N,C,X
     GO TO 999
 360 TYPE 1001,N2,A,N1,D,N2,B,N1,C,X
     GO TO 999
2000 IF (E1-E2)166,260,366
 166 TYPE 6040
     ACCEPT 6000,MM
     IF (MM) 168,165,168
 168 TYPE 6040
     ACCEPT 6000,KK
     IF (KK) 185,185,887
 187 TYPE 4002,N2,A,D,N1,C,B,X
4002 FORMAT (//1X,I2,A4,' + 'A4,'='I2,A4,' + 'A4,2X,'CELL EMF = '
    +F6.3,' VOLTS'/)
     GO TO 999
 185 TYPE 3001,C,B,D,N2,A,X
3001 FORMAT (//A4,' + 'A4,'='A4,' + 'I2,A4/
    1'  CELL EMF = 'F6.3,' VOLTS ')
     GO TO 999
 165 TYPE 1002,C,B,A,D,X
1002 FORMAT (//1X,A4,' + 'A4,' = 'A4,' + 'A4, 5X, 'CELL EMF
    +='F6.3,' VOLTS')
     GO TO 999
 366 TYPE 6040
     ACCEPT 6000,JJ
     IF (JJ) 365,365,368
 368 TYPE 6040
     ACCEPT 6000,II
     IF (II) 885,885,187
 885 TYPE 3001,A,D,B,N2,C,X
     GO TO 999
 887 TYPE 4002,N2,C,B,N1,A,D,X
     GO TO 999
 365 TYPE 1002,A,D,B,C,X
     GO TO 999
 260 TYPE 6060
 999 TYPE 6070
6070 FORMAT ( ' TO PROCEED - TYPE 1. ',$)
6060 FORMAT (' NO REACTION; EO''S EQUAL.')
     ACCEPT 6000,K
6000 FORMAT (G)
5001 CONTINUE
     END
```

Fig. 15. — Continued.

```
IN THIS PROGRAM , FOLLOWING OXIDATION POTENTIAL
CONVENTION IS USED:
RED.FORM = OXD.FORM + (N)E-   EO
 A = B + (N1)E-     E1
 C = D + (N2)E-     E2


THIS PROGRAM AS DESIGNED WILL NOT TAKE
CARE OF THE HALF REACTION

H2 = 2H+ + 2E- EO = 0.0

DO NOT USE THE ABOVE MENTIONED
REACTION AS ONE OF YOUR INPUT.


APPROXIMATELY HOW MANY TIMES
WILL YOU USE THIS PROGRAM 3

WHAT ARE YOUR A, B, N1 AND E1?
K   K+1 1,+2.925

WHAT ARE YOUR C, D, N2 AND E2?
AU  AU+33,-1.50

IS ONE OF THE SPECIES DIATOMIC? YES(1),NO(0). 0


3K     +  1AU+3 =  3K+1   +  1AU      CELL EMF =  4.425 VOLTS
TO PROCEED - TYPE 1. 1

WHAT ARE YOUR A, B, N1 AND E1?
NI   NI+22,.250

WHAT ARE YOUR C, D, N2 AND E2?
CL-1CL2 2,-1.359

IS ONE OF THE SPECIES DIATOMIC? YES(1),NO(0). 1

IS ONE OF THE SPECIES DIATOMIC? YES(1),NO(0). 0


I    + CL2 =NI+2 +  2CL-1
 CELL EMF =  1.609 VOLTS
TO PROCEED - TYPE 1. 1

WHAT ARE YOUR A, B, N1 AND E1?
CU  CU+11,-.521

WHAT ARE YOUR C, D, N2 AND E2?
CU  CU+22,-.337

IS ONE OF THE SPECIES DIATOMIC? YES(1),NO(0). 0


1CU    +  2CU+1 =  2CU    +  1CU+2   CELL EMF =  0.184 VOLTS
```

Fig. 16. Sample output from program EO.

4. THERMO

This program is designed to calculate a full range of thermodynamic quantities — ΔH, ΔE, ΔS, and ΔG — based on an appropriate sequence of standard input parameters. A listing of the complete FORTRAN program is given in Fig. 17. Following a specification of the reaction itself, the user is lead through a series of questions calling for standard enthalpies of formation and absolute entropies as needed (see Fig. 18). The final ΔG value is derived from the calculated ΔH and ΔS quantities. This program was particularly effective at stimulating discussion about the relative magnitudes and sign of ΔE and ΔH, as well as the effect of ΔS on the sign of the free energy change for the reaction. The calculation of the equilibrium constant for the reaction also served to link the topic of thermodynamics to the concept of equilibrium.

```
THERMO


0
10  PRINT "THIS PROGRAM IS DESIGNED FOR"
20  PRINT "THE FOLLOWING TYPE OF EQUATION:"
25  PRINT
30  PRINT "A+B+C=D+E+F"
35  PRINT
40  PRINT "HOW MANY TIMES WILL YOU USE "
42  PRINT "THIS PROGRAM";
45  INPUT N1
50  FOR I1=1 TO N1
55  PRINT "INPUT THE REACTION"
60  INPUT A$
70  PRINT "WHAT ARE THE COEFFICIENTS OF"
71  PRINT "REACTANTS";
80  INPUT R1,R2,R3
90  PRINT "WHAT ARE THE COEFFICIENTS OF"
91  PRINT "PRODUCTS";
100 INPUT R4,R5,R6
120 PRINT " IS DELTA-E KNOWN? YES OR NO"
130 INPUT I$
140 IF I$="NO" THEN 210
150 PRINT "WHAT ARE  DELTA-E,DELTA-N,"
151 PRINT"R AND T"
160 INPUT D1,D2,R,T
170 W=(D2*R*T)/1000.0
180 D3=D1+W
190 GO TO 310
210 PRINT "WHAT ARE THE HEATS OF"
211 PRINT "FORMATION FOR THE REACTANTS"
220 INPUT H1,H2,H3
230 PRINT "WHAT ARE THE HEATS OF"
231 PRINT "FORMATION FOR THE PRODUCTS"
240 INPUT H4,H5,H6
250 D3=(R4*H4+R5*H5+R6*H6)-(R1*H1+R2*H2+R3*H3)
```

Fig. 17. FORTRAN listing of program THERMO.

```
260 PRINT "WHAT ARE DELTA-N,R AND T"
270 INPUT D2,R,T
280 W=(D2*R*T)/1000.0
290 D1=D3-W
310 PRINT "IS DELTA-S KNOWN? YES OR NO"
320 INPUT J$
330 IF J$="YES" THEN 400
340 PRINT "WHAT ARE THE ABSOLUTE"
341 PRINT "ENTROPIES FOR THE REACTANTS"
350 INPUT S1,S2,S3
360 PRINT "WHAT ARE THE ABSOLUTE"
361 PRINT "ENTROPIES OF THE PRODUCTS"
370 INPUT S4,S5,S6
380 D8=(R4*S4+R5*S5+R6*S6)-(R1*S1+R2*S2+R3*S3)
390 D5=D3-(T*D8/1000.)
395 GO TO 420
400 PRINT"WHAT IS DELTA-S"
410 INPUT D8
420 E1=10.0**-(D5*1000.0/(2.303*R*T))
440 PRINT A$
450 PRINT USING 460,D3,W
460 :DH= ###.###      W= ###.###
465 PRINT USING 466,D1
466 :DE= ###.###
467 PRINT
470 PRINT USING 480,D8,D5
480 :DS= ###.###      DG= ###.###
481 PRINT USING 482,E1
482 :K-EQ= ##.####↑↑↑↑
484 PRINT "TO PROCEED TYPE C"
486 INPUT K$
488 NEXT I1
490 END
```

Fig. 17.— Continued.

5. ISOMER

This program is directed toward a topic often covered in the elementary treatment of organic chemistry and now included in many general chemistry courses. The program (see Fig. 19 for listing) calculates the numbers of primary, secondary, and tertiary isomers which are possible for mono-functional alcohols containing no more than 22 carbon atoms [2]. There are two forms for the output (see Figs. 20 and 21): one abbreviated version which is suitable for the 32-character linewidth used for video display, and one expanded version which is optionally available when video display is not being used. Certainly this program involves calculations which are much too involved to perform manually in real time during a lecture. Beyond six — or seven — carbon alcohols the number of isomers is so large that there is little or no chance, however, of developing class discussion around the various isomeric structures.

```
THERMO


THIS PROGRAM IS DESIGNED FOR
THE FOLLOWING TYPE OF EQUATION:

A+B+C=D+E+F

HOW MANY TIMES WILL YOU USE
THIS PROGRAM ?2
INPUT THE REACTION
 ?FE2O3(S)+3H2(G)=2FE(S)+3H2O(L)
WHAT ARE THE COEFFICIENTS OF
REACTANTS ?1,3,0
WHAT ARE THE COEFFICIENTS OF
PRODUCTS ?2,3,0
 IS DELTA-E KNOWN? YES OR NO
 ?NO
WHAT ARE THE HEATS OF
FORMATION FOR THE REACTANTS
 ?-196.5,0.0,0.0
WHAT ARE THE HEATS OF
FORMATION FOR THE PRODUCTS
 ?0.0,-68.32,0.0
WHAT ARE DELTA-N,R AND T
 ?-3.0,1.987,298.15
IS DELTA-S KNOWN? YES OR NO
 ?NO
WHAT ARE THE ABSOLUTE
ENTROPIES FOR THE REACTANTS
 ?21.5,31.21,0.0
WHAT ARE THE ABSOLUTE
ENTROPIES OF THE PRODUCTS
 ?6.49,16.73,0.0
FE2O3(S)+3H2(G)=2FE(S)+3H2O(L)
DH=  -8.460      W=  -1.777
DE=  -6.683

DS= -51.960      DG=   7.032
K-EQ=  7.015E-06
TO PROCEED TYPE C
 ?C
```

```
INPUT THE REACTION
 ?H2(G)+CL2(G)=2HCL(G)
WHAT ARE THE COEFFICIENTS OF
REACTANTS ?1,1,0
WHAT ARE THE COEFFICIENTS OF
PRODUCTS ?2,0,0
 IS DELTA-E KNOWN? YES OR NO
 ?YES
WHAT ARE  DELTA-E,DELTA-N,
R AND T
 ?-44.12,0.0,1.987,298.15
IS DELTA-S KNOWN? YES OR NO
 ?NO
WHAT ARE THE ABSOLUTE
ENTROPIES FOR THE REACTANTS
 ?31.21,53.3,0.0
WHAT ARE THE ABSOLUTE
ENTROPIES OF THE PRODUCTS
 ?44.6,0.0,0.0
H2(G)+CL2(G)=2HCL(G)
DH= -44.120      W=   0.000
DE= -44.120

DS=   4.690      DG= -45.518
K-EQ=  2.305E+33
TO PROCEED TYPE C
 ?C
```

Fig. 18. Sample output from program THERMO.

```
C THIS PROGRAM CALCULATES THE NUMBER OF STRUCTURAL ISOMERS
C FOR ANY MONOFUNCTIONAL ALCOHOL HAVING LESS THAN 100 C ATOMS.
C WRITTEN BY J. W. MOORE, FEBRUARY, 1972.
C SEE ARTICLE IN 'CHEMISTRY' FEB., 1972, P. 6, FOR DETAILS.
      DIMENSION NT(100)
      REAL NT,NP,NS,NTA,NTB,NTC,NTOTT
      TYPE 200
200   FORMAT(' THIS PROGRAM CALCULATES THE',/
     1' NUMBER AND TYPE OF STRUCTURAL',/
     2' ISOMERS FOR MONOFUNCTIONAL',/
     3' ALCOHOLS UP TO 22 CARBON ATOMS.',/
     4' IF USING TV DISPLAY TYPE TV',/
     5' IF NOT TYPE TTY.   ',$)
      ACCEPT 210,I23
210   FORMAT(A2)
      ITV=1
1     TYPE 220
220   FORMAT(' TYPE NUMBER OF CARBONS.   ',$)
      ACCEPT 230,N
230   FORMAT(G)
      IF(I23 .EQ. 'TV   ') ITV=0
      IF(N)99,99,2
2     NN=N-1
      NT(1) = 1
      DO 3  I=1,NN
      NTA=0
      NTB=0
      NTC=0
      NS=0
      NP=0
      NP = NT(I)
      IF(I-1) 4,4,5
5     AI=I
      IEVEN=AI/2.0 + 0.51
      IF(IEVEN-I/2) 6,6,7
6     I02M1=I/2 - 1
      DO 60  J=1,I02M1
60    NS=NS+NT(J)*NT(I-J)
      IF(I-2)8,8,9
9     NS=NS+0.5*NT(I/2)*(1+NT(I/2))
      GO TO 8
7     IM102=(I-1)/2
      DO 70  J=1,IM102
70    NS=NS+NT(J)*NT(I-J)
8     IF(I-2) 4,4,12
12    IF(I-5)13,13,14
14    IM3=I-3
      DO 15 J=1,IM3
      JP1=J+1
      DO 15 K=JP1,IM3
      L=I-(J+K)
      IF(L-K) 15,15,16
16    NTA=NT(L)*NT(K)*NT(J) + NTA
15    CONTINUE
```

Fig. 19. FORTRAN listing of program ISOMER.

```
13     IF(I-3) 17,17,18
18     IF(IEVEN-I/2) 19,19,20
19     JJ=2
       GO TO 21
20     JJ=1
21     IM2=I-2
       DO 190 J=JJ,IM2,2
       K=(I-J)/2
       IF(K-J)191,190,191
191    NTB=NTB+0.5*NT(K)*(1+NT(K))*NT(J)
190    CONTINUE
17     AI=I
       I3=AI/3.0+0.78
       IF(I3-I/3)22,22,4
22     NTC=NT(I/3)*(1+NT(I/3))*(2+NT(I/3))/6
4      NT(I+1)=NP +NS+NTA+NTB+NTC
3      CONTINUE
       TYPE 100,N
100    FORMAT(' C ',I2,' ALCOHOLS HAVE')
       TYPE 101, NT(N)
101    FORMAT(F12.0,' STRUCTURAL ISOMERS:')
       TYPE 102,NP
102    FORMAT('      ',F12.0,'-PRIMARY,')
       TYPE 103,NS
103    FORMAT(6X,F12.0,'-SECONDARY,')
       NTOTT=NTA+NTB+NTC
       TYPE 107,NTOTT
107    FORMAT(6X,F12.0,'-TERTIARY.')
       IF(ITV)97,97,98
98     TYPE 104,NTA
104    FORMAT(' SUBTOTAL OF TERT ISOMERS, R(1) UNEQ R(2) UNEQ ',
      1  'R(3), IS',F12.0)
       TYPE 105,NTB
105    FORMAT(' SUBTOTAL OF TERT ISOMERS, R(1) UNEQ R(2) EQ ',
      1  'R(3), IS',F14.0)
       TYPE 106,NTC
106    FORMAT(' SUBTOTAL OF TERT ISOMERS, R(1) EQ R(2) EQ ',
      1   'R(3), IS',F16.0)
97     CONTINUE
       TYPE 240
240    FORMAT(' DONE? TYPE A ZERO.',/
      1' MORE ALCOHOLS?')
       GO TO 1
99     CONTINUE
       STOP
       END
```

Fig. 19. — Continued.

```
THIS PROGRAM CALCULATES THE
NUMBER AND TYPE OF STRUCTURAL
ISOMERS FOR MONOFUNCTIONAL
ALCOHOLS UP TO 22 CARBON ATOMS.
IF USING TV DISPLAY TYPE TV
IF NOT TYPE TTY.  TV

TYPE NUMBER OF CARBONS.  4

C  4 ALCOHOLS HAVE
         4. STRUCTURAL ISOMERS:
               2.-PRIMARY,
               1.-SECONDARY,
               1.-TERTIARY.
DONE? TYPE A ZERO.
MORE ALCOHOLS?
TYPE NUMBER OF CARBONS.  8

C  8 ALCOHOLS HAVE
        89. STRUCTURAL ISOMERS:
              39.-PRIMARY,
              33.-SECONDARY,
              17.-TERTIARY.
DONE? TYPE A ZERO.
MORE ALCOHOLS?
TYPE NUMBER OF CARBONS.  16

C 16 ALCOHOLS HAVE
    124906. STRUCTURAL ISOMERS:
            48865.-PRIMARY,
            49060.-SECONDARY,
            26981.-TERTIARY.
DONE? TYPE A ZERO.
MORE ALCOHOLS?
TYPE NUMBER OF CARBONS.  0
```

Fig. 20. Video output from program ISOMER.

A number of other programs have been used for on-line classroom computing in general chemistry; however, they will not be individually discussed in detail since the previous set of programs illustrates adequately the pedagogical approach involved. The other topics include: predicting the electron configurations of atoms and ions in s, p, d terminology; comparing ideal vs van der Waals gas law calculations (Figs. 22 and 23); solving stoichiometry problems; predicting molecular geometry via the Valence Shell Electron Pair Repulsion (VSEPR) theory; calculating solution concentrations and the resultant colligative properties; and performing calculations based on the Bohr model of the hydrogen atom. In addition, a variety of programs have been developed for use in specialized advanced courses such as x-ray crystallography.

```
THIS PROGRAM CALCULATES THE
NUMBER AND TYPE OF STRUCTURAL
ISOMERS FOR MONOFUNCTIONAL
ALCOHOLS UP TO 22 CARBON ATOMS.
IF USING TV DISPLAY TYPE TV
IF NOT TYPE TTY.  TTY

TYPE NUMBER OF CARBONS.  4

C  4 ALCOHOLS HAVE
         4. STRUCTURAL ISOMERS:
              2.-PRIMARY,
              1.-SECONDARY,
              1.-TERTIARY.
SUBTOTAL OF TERT ISOMERS, R(1) UNEQ R(2) UNEQ R(3), IS            0.
SUBTOTAL OF TERT ISOMERS, R(1) UNEQ R(2) EQ R(3), IS              0.
SUBTOTAL OF TERT ISOMERS, R(1) EQ R(2) EQ R(3), IS                1.
DONE? TYPE A ZERO.
MORE ALCOHOLS?
TYPE NUMBER OF CARBONS.  8

C  8 ALCOHOLS HAVE
        89. STRUCTURAL ISOMERS:
             39.-PRIMARY,
             33.-SECONDARY,
             17.-TERTIARY.
SUBTOTAL OF TERT ISOMERS, R(1) UNEQ R(2) UNEQ R(3), IS            4.
SUBTOTAL OF TERT ISOMERS, R(1) UNEQ R(2) EQ R(3), IS             13.
SUBTOTAL OF TERT ISOMERS, R(1) EQ R(2) EQ R(3), IS                0.
DONE? TYPE A ZERO.
MORE ALCOHOLS?
TYPE NUMBER OF CARBONS.  16

C 16 ALCOHOLS HAVE
    124906. STRUCTURAL ISOMERS:
           48865.-PRIMARY,
           49060.-SECONDARY,
           26981.-TERTIARY.
SUBTOTAL OF TERT ISOMERS, R(1) UNEQ R(2) UNEQ R(3), IS        15875.
SUBTOTAL OF TERT ISOMERS, R(1) UNEQ R(2) EQ R(3), IS          10986.
SUBTOTAL OF TERT ISOMERS, R(1) EQ R(2) EQ R(3), IS              120.
DONE? TYPE A ZERO.
MORE ALCOHOLS?
TYPE NUMBER OF CARBONS.  0
```

Fig. 21. Non-video output from program ISOMER.

VI. SUMMARY

On-line classroom computing with accompanying video display of the output appears to be a valuable teaching aid. It not only makes available during the lecture the full computational power of the computer, but it dramatically enlivens the entire classroom environment. The technique is virtually unrivaled for stimulating student participation and promoting active discussion. The implementation of the method via the use of the compact, inexpensive video adapter is particularly convenient and, based on experience to date, trouble-free. The exact pedagogical strategy to be used for optimizing the impact of on-line classroom computing on student learning is, of course, subject to discussion. For example, the relative merits of

```
VANDER

1 'THIS PROGRAM WAS DEVELOPED BY DR. C. T. FURSE AS AN'
2 'ON LINE CLASSROOM DEMONSTRATION OF THE ITERATION TECHNIQUE'
3 'IN THE SOLVING OF THE QUADRATIC EQUATION, AND AS A '
4 'COMPARISION OF THE IDEAL GAS LAW VOLUMES AND VAN DER'
10 'WAALS VOLUMES.'
20 DIMENSION GAS(2)
30 PRINT "ENTER VALUE OF GAS LAW"
40 PRINT "CONSTANT (R)"
50 INPUT,R
60 10 PRINT "NAME OF GAS"
70 INPUT 9,GAS
80 PRINT 19,GAS
90 INPUT,A,B
100 PRINT "ENTER TEMPERATURE, NUMBER"
110 PRINT "OF MOLES, STARTING PRESSURE"
120 PRINT "INCREMENT OF PRESSURE"
130 PRINT "AND MAXIMUM PRESSURE"
140 INPUT,T,SN,P,DELP,PMAX
150 C=0.0001
160 MNI=150
170 N=1
175 PRINT "DO YOU WANT TO ILLUSTRATE THE"
176 PRINT "ITERATION TECHNIQUE, TYPE"
177 PRINT "YES OR NO"
178 INPUT 79,IJK
179 IF(IJK-"YES")200,500,200
180 200 PRINT "      YDEAL    VAN DER WAALS"
190 PRINT "  P        V           V"
210 400 V=SN*R*T/P
220 VI=V
230 PI=P
240 1808 VW=(SN*R*T)/(P+A*SN*SN/(V*V))+SN*B
250 D=ABS(VW-V)
260 V=VW
270 N=N+1
280 IF (D-C)999,999,1003
290 1003 IF (N-MNI)1808,1808,1010
300 1010 PRINT"NO CONVERGENCE"
310 GO TO 10
```

Fig. 22. FORTRAN listing of program VANDER.

```
320 999 PRINT 39,PI,VI,V
330 P=P+DELP
335 N=1
340 IF(P-PMAX)400,400,1800
350 1800 GO TO 10
360 9 FORMAT(2A3)
370 19 FORMAT(1X,"VAN DER WAALS CONSTANTS"/
380 +"A AND B FOR ",2A3," ARE"
390 +)
400 39 FORMAT(1X,F5.2,3X,F8.4,3X,F8.4)
500 500 V=SN*R*T/P
510 VI=V
520 PI=P
530 700 VW=(SN*R*T)/(P+A*SN*SN/(V*V))+SN*B
540 D=ABS(VW-V)
550 V=VW
560 PRINT 59,N,V
570 N=N+1
580 IF(D-C)800,800,810
590 810 IF(N- 20)700,700,830
600 830 PRINT "NO CONVERGENCE"
610 800 PRINT 69,PI,VI,V
620 N=1
630 PRINT "DO YOU WANT TO CONTINUE"
631 INPUT 79,IYES
632 79 FORMAT(2A3)
633 IF(IYES-"YES")10,200,10
650 59 FORMAT("ITERATION ",I2,2X,"VOLUME ",F8.4)
660 69 FORMAT("P = ",F5.2,2X,"IDEAL V = ",F8.4/
670 +"VAN DER WAALS V = ",F8.4)
1000 END
```

Fig. 22. — Continued.

extending the technique into the grade structure via the use of computer-oriented homework assignments based on utilization of previously demonstrated programs can be argued. Also, the use of the computer in this mode to perform relatively trivial computations simply for the sake of rapidity and extended accuracy can also be debated. Nevertheless, based on preliminary studies the bulk of the evidence seems to suggest that on-line classroom computing is quite effective instructionally.

This conclusion is derived from the enthusiastic comments which students made in response to a course evaluation question concerning what contribution if any on-line classroom computing made to their understanding of the course material. Furthermore, from the instructor's perspective, the use of more sophisticated and more accurate methods in class without excessive loss of time should produce a better lecture coverage of quantitative aspects of chemistry. No attempts have been made, however, to study the impact of this technique by the use of appropriate control groups and comparative student performances. Certainly parameters such as class size, the number of video monitors, and the frequency of use of classroom computing are important; nevertheless, the excitement and impact of this method can be felt without elaborate educational testing. The question of the cost effectiveness of on-line classroom computing is also an obvious consideration, but if the frequency of use is limited to approximately once per week, the cost would appear to be in the range of $5-$10 per classroom period. This estimate is subject to wide local variation and is, of course, highly

```
VANDER


ENTER VALUE OF GAS LAW
CONSTANT (R)
? .0821
NAME OF GAS
? O2
 VAN DER WAALS CONSTANTS
A AND B FOR O2      ARE
? 1.36, .0318
ENTER TEMPERATURE, NUMBER
OF MOLES, STARTING PRESSURE
INCREMENT OF PRESSURE
AND MAXIMUM PRESSURE
? 273,1,1,2,12
DO YOU WANT TO ILLUSTRATE THE
ITERATION TECHNIQUE, TYPE
YES OR NO
? NO
     IDEAL    VAN DER WAALS
  P       V          V
  1.00     22.4133     22.3844
  3.00      7.4711      7.4422
  5.00      4.4827      4.4538
  7.00      3.2019      3.1731
  9.00      2.4904      2.4616
  11.00      2.0376      2.0088
NAME OF GAS
? O2
 VAN DER WAALS CONSTANTS
A AND B FOR O2      ARE
? 1.36,0.0318
ENTER TEMPERATURE, NUMBER
OF MOLES, STARTING PRESSURE
INCREMENT OF PRESSURE
AND MAXIMUM PRESSURE
? 103,1,5,3,20
DO YOU WANT TO ILLUSTRATE THE
ITERATION TECHNIQUE, TYPE
YES OR NO
? NO
     IDEAL    VAN DER WAALS
  P       V          V
  5.00      1.6913      1.5513
  8.00      1.0570       .9083
  11.00      .7688       .6078
  14.00      .6040       .4239
  17.00      .4974       .2528
  20.00      .4228       .0431
```

Fig. 23. Sample output from program VANDER.

```
NAME OF GAS
? 02
 VAN DER WAALS CONSTANTS
A AND B FOR 02      ARE
? 1.36,0.0318
ENTER TEMPERATURE, NUMBER
OF MOLES, STARTING PRESSURE
INCREMENT OF PRESSURE
AND MAXIMUM PRESSURE
? 103,1,17,2,25
DO YOU WANT TO ILLUSTRATE THE
ITERATION TECHNIQUE, TYPE
YES OR NO
? YES
ITERATION  1  VOLUME     .4077
ITERATION  2  VOLUME     .3676
ITERATION  3  VOLUME     .3443
ITERATION  4  VOLUME     .3288
ITERATION  5  VOLUME     .3177
ITERATION  6  VOLUME     .3093
ITERATION  7  VOLUME     .3027
ITERATION  8  VOLUME     .2973
ITERATION  9  VOLUME     .2929
ITERATION 10  VOLUME     .2892
ITERATION 11  VOLUME     .2861
ITERATION 12  VOLUME     .2833
ITERATION 13  VOLUME     .2809
ITERATION 14  VOLUME     .2788
ITERATION 15  VOLUME     .2770
ITERATION 16  VOLUME     .2753
ITERATION 17  VOLUME     .2738
ITERATION 18  VOLUME     .2724
ITERATION 19  VOLUME     .2712
ITERATION 20  VOLUME     .2701
NO CONVERGENCE
P =  17.00  IDEAL V =     .4974
VAN DER WAALS V =    .2701
DO YOU WANT TO CONTINUE
? YES
     IDEAL     VAN DER WAALS
  P       V          V
  17.00      .4974        .2528
  19.00      .4451        .0431
  21.00      .4027        .0430
  23.00      .3677        .0430
  25.00      .3383        .0429
```

Fig. 23.— Continued.

dependent on the complexity of the programs used; however, the estimate is accurate enough to indicate that the cost is not prohibitive. The cost is extremely minimal if calculated on a per student basis for the entire class. Overall, on-line classroom computing overcomes some of the factors which have been cited [3] as tending to inhibit the use of computers in instruction.

ACKNOWLEDGMENTS

I would like to express my thanks to my colleagues at Eastern Michigan University, J. W. Moore, C. T. Furse, and S. K. Vaidya, for their assistance in the preparation of many of the programs described in this chapter.

REFERENCES

1. R. W. Collins, Proceedings of the Second Conference on Computers in the Undergraduate Curricula, Compute, Hanover, New Hampshire, 1971, p. 99.

2. D. W. Rouvray, Chemistry, 45 (2), 6 (1972).

3. E. J. Anastasio and J. S. Morgan, Factors Inhibiting the Use of Computers in Instruction, EDUCOM Report, 1972.

Chapter 5

INFORMATION STORAGE AND RETRIEVAL FOR CHEMISTS: COMPUTERIZED SYSTEMS, SOURCES, AND SERVICES

Martha E. Williams

Coordinated Science Laboratory
University of Illinois
Urbana, Illinois

I. INTRODUCTION

A. Chemical Information

Chemical information and data are the products of testing, research, and development, and they become a resource for new or further research and development. Most chemists have an appreciation for, or value, information and data. They are also becoming increasingly aware of the fact that today information can be generated, processed, stored, managed, retrieved, and used in new ways. This is a result of the increased economical use of computers in chemistry. Computers are used as tools in research. They are used for control and monitoring of experiments, and for simulating experiments — whether it be for analysis, kinetics, or synthesis of compounds. Once the research or testing is completed the products of the work, i.e., the data, can be selected and reduced from analog to digital form. These computer-generated data — in either human-readable or machine-readable form, on paper, magnetic tape, disks, drums, chips, data cells, or microform — now become a part of a vast, varied, and disperse data reservoir.

Information and data can be considered to be of several types: (1) digitized numeric, as in the case of chemical handbooks, (2) analog, as in the case of IR or other spectral data, (3) symbolic, as in the case of structures of chemical compounds, and (4) alphabetic, as in the case of information expressed in natural language. These data, if they are to be used again as an input to further research, must be stored on an appropriate medium, arranged in a file structure suitable for searching, and managed intelligently. The existence of such data bases together with search systems enables us to conduct searches and retrieve information for users. The researcher uses the system to obtain background data, keep up-to-date in his field, supplement the information he has, identify persons, organizations, documents, or other sources of information, or to make judgments as to whether the research he proposes has already been done.

Machine-readable data bases have been developed as direct products of research and as secondary or tertiary products. That is, research may be directed toward generation of data — IR spectra for example — or, the results of research may be written up in journal articles that are then abstracted and indexed in secondary sources. The journal information or the abstract/index information may be put in machine-readable form either for purposes of searching or for purposes of automating production and printing functions. Prior to the advent of economic computer composition the cost of keypunching data to create large files for computer searching was very high and could seldom be justified on the basis of information retrieval alone.

Now that large data bases are often created as a by-product of publication, the cost factor for input is no longer so significant. Many new data bases have become publicly available from commercial organizations, government agencies, and scientific and technical societies; it behooves the chemist to become aware of these new sources of chemical information and to learn how to use them. He may use them within his own organization or he may purchase the service from information brokers in information centers.

B. History and Growth of Chemical Literature

Chemists are probably more fortunate than other scientists in having access to the literature of their field, because chemistry has a tradition of information excellence and information organization. Chemists have been concerned with the problems of cataloging, indexing, abstracting, storing, and retrieving information for a long time.

The first journal devoted to chemistry, Crell's Chemisches Journal für die Freunde der Naturleben, appeared in 1778. In 1821 the volume of chemical literature had reached the stage where Berzelius began his series of chemical reviews, "Jahresberichte über die Fortschritte der physischen Wissenschaften." In 1907 the first volume of Chemical Abstracts was published; it has grown to the point where in 1972 it published 334,426 abstracts of chemical papers and patents appearing in 12,000 scientific and technical and trade journals published in 106 nations in 56 languages.

Beilstein once said, "I read everything, I place it where it belongs." While in past years it was possible for a scientist to keep up in his area of specialization and be aware of relevant current developments by scanning a couple hundred papers a year, this is no longer possible (at least for the majority) because the volume of literature one must scan has become so large and because the subject matter of scientific disciplines is no longer as clearly demarcated as in the past — much of the work that is done is interdisciplinary in character. Evidence of the exponential growth of the chemical literature is the fact that it took 32 years for CA to publish its first million abstracts, 18 years the second million, eight years the third, four years eight months for the fourth, and three years four months for the fifth million. Today some 100,000 scientific journals provide channels of communication for the world's scientific knowledge. How is the individual scientist to glean the material he needs from this vast store? In recent years secondary publications (abstracting, indexing, and alerting journals) such as Chemical Abstracts, Chemical Titles, and Current

Contents were used to provide a first screening or winnowing of information by making broad subspecialty groupings of categories and abstracts. These are no longer sufficient; newer computer-based search services have come into existence to provide rapid access to scientific and technical literature.

II. CONTENT ANALYSIS AND VOCABULARY CONTROL FOR SEARCH SYSTEMS

A. Introduction

Information systems involve the documents, articles, reports or other source items that make up the data bank, and a means of retrieving the items from the bank. These items must be stored on some type of physical storage media (hard copy, microform, magnetic tape, etc.). In order to locate documents or other items pertinent to a search question, the documents must be uniquely addressable or identifiable and they should be accessible by subject content. Thus, it is necessary to associate subject categories, classes, index terms or other labels or tags with each item in a file in order to retrieve it in a meaningful way.

Assuming that pertinent items have been selected for inclusion in an information system, indexing and classification for subsequent retrieval are the most important tasks involved in establishing information systems—libraries, specialized information centers, data analysis centers or data banks. Classification involves distribution into groups and the systematic division of a group of related subjects. A classification system is a schedule for arranging or organizing documents. Classes can group items by subject or by property, e.g., size, language, date, alphabet, or form. An index is a file or list of labels or tags for documents or data. It is a systematic or organized list that indicates the information content of the item and relates this by means of labels to the storage location of the item.

The creation of these labels—the process of indexing—involves analyzing the document containing the information and formatting the results of the analysis in an indexing language. The creation of the indexing language involves both generation and control of the indexing terms.

The types of search systems in use range from conventional library card catalogs to sophisticated computer systems; they are distinguished by type of recording media on which the information store and the index to the store are maintained or manipulated, and by the type of equipment used for searching the recorded information. The principal types of search systems include computers, microform selection and scanning equipment, EAM cards (electronic counting machine cards) and their associated unit record equipment, edge-notched punch card sorting equipment, and optical coincidence equipment.

Retrieval systems may retrieve document numbers or actual data. In most cases, index terms are stored for searching and retrieval, and the use of mechanical or electronic equipment is largely for searching indexes to determine which documents contain desired information. An information retrieval system can also search full text files or data files; in these cases the system can truly be referred to as an information retrieval system. Or, as is more frequently the case, the system may retrieve document surrogates — abstracts; digests; extracts; and citations or bibliographic references with source identification, e.g., author and title; document numbers or document addresses. With the exception of direct searching of full text or data files, an information system accomplishes the retrieval by searching an index to the collection. The indexes exist because the original volume of information is too large to be searched directly.

In order to make best use of the varied chemical information sources and services, it is necessary to understand the basic indexing, classification, and control principles that govern the data input. Different sources employ different indexing schemes, index to different depths, and use varied nomenclature. There are both traditional and nonconventional systems and techniques.

Traditional library classification schemes include the Dewey Decimal Classification (DDC), the Library of Congress (LC), and the Universal Decimal Classification (UDC) systems. In these cases subject and bibliographic information used for searching is maintained on library cards. (In 1969, the Library of Congress intitiated the MARC II service for providing catalog information on magnetic tapes.) Nonconventional retrieval systems employ different methods of indexing and classifying information, and different techniques for recording, storing, and manipulating index records. Techniques employed in nontraditional information systems may involve uniterms, concept coordination, superimposed coding, descriptors, semantic coding, and links and role indicators. In these cases subject and bibliographic information (and even abstracts) can be maintained on conventional file cards, edge-notched punch cards, or optical coincidence cards. Nonconventional systems involve the development of controlled vocabularies, authority lists, and thesauri for preparation of index term input. These and other word guides, dictionaries, and term frequency lists are employed as user aids in preparing search questions and interest profiles.

B. Indexing Systems and Techniques

Classification systems that attempt to organize all knowledge in a highly structured hierarchy are not particularly suited to accommodate the expanding scope of scientific and technical knowledge and the changing and variable interpretations of this knowledge. The growth of scientific and technical publications together with increased complexity and detail of information has forced the development of newer methods of indexing and classifying and new techniques for recording and manipulating index terms. The use of controlled vocabularies with cross references to broader, narrower, synonymous and related terms, and the further use of links and roles are all classificatory procedures. They imply relations and add specificity to indexing systems.

Mortimer Taube developed a system of "uniterm" indexing and the concept of "coordinate indexing". Calvin Mooers was responsible for the development of "superimposed coding" and the use of "descriptors." James Perry and Allen Kent at Western Reserve University developed a system of "semantic coding" employing "role indicators" that laid the groundwork for link and role indexing.

1. Coordinate Indexing

The "uniterm" system of "coordinate indexing" was developed to overcome the difficulty encountered in library subject catalogs of presenting subject headings according to a specific permuted order. For example, a book on Qualitative Analysis might be cataloged under the subject heading "Chemistry, Analytic-Qualitative". The three subject terms are listed in a specific permuted order and the book would not be found by looking under the terms "Qualitative Analysis" or "Analytic Chemistry". The use of coordinate indexing permits coordination of terms or concepts by combination rather than permutation. Uniterms are single words or concepts that describe document contents. In a coordinate indexing scheme each single concept, subject, or uniterm is assigned to a card and the identification numbers of all documents that have been indexed by that term are recorded on the card. When information combining two or more subjects is desired, a coordination search is conducted by examining the appropriate term cards and comparing the document numbers listed on each card. The basic purpose of concept coordination indexing is to avoid precoordination of terms by using single concepts or single terms.

Coordinate indexing does solve the problem that occurs when several concepts (e.g., chemistry, analysis, quality) that are combined to make up a subject index term are listed in only one of the several possible permuted orders. Coordinate indexing does not, however, solve the problem of incorrect relationships between terms. For example, a document dealing with "principles of formulation in chemistry" would be indexed under each of the three terms. By coordinating the three terms this document would be properly retrieved by a user looking for information on principles of formulation in chemistry, but it would also be erroneously retrieved by a user looking for information on the formulation of principles in chemistry. In this case it would constitute a false drop or false retrieval.

While in the original system the uniterms were recorded on cards, the principle of concept coordination has been used in systems that employ other media for recording and searching. Concept coordination is used in optical coincidence systems and in computer inverted files. In an optical coincidence system the document numbers are recorded as drilled, dedicated spaces on a term card. A search coordinating several terms or concepts is conducted by placing the appropriate term cards over a light source. Light will shine through only where the same document number space has been drilled for all the cards. The principle of concept coordination has been used, for example, by the American Society for Testing and Materials (ASTM) for their infrared spectral (IR) data index, X-ray powder diffraction data, nuclear magnetic resonance (NMR) data, and mass spectroscopy (MS) data. These data systems are maintained on optical coincidence cards.

The IR system is used to identify unknown compounds on the basis of IR spectra. In the IR system, optical coincidence cards represent:

Deck A. (1) Peaks present + 0.1 μ (nearest odd 0.1 μ)
(2) Peaks absent 0.3 μ range

Deck B. (1) Sample state (solid, liquid, gas)
(2) Elements present and absent
(3) Functional groups present and absent

Deck C. (1) Peaks present + 0.1 μ (nearest even 0.1 μ)

Card deck A is the basic component of the index and provides for searches by the presence or absence of peaks. Card deck B adds the ability to search by physical state of the sample, by the presence or absence of oxygen, nitrogen, sulfur, and the halogens, and by the presence or absence of the following functional groups:

−OH	includes Hydroxyl
−O−	includes Ether
>C=O	includes Carbonyl
−N<	includes Amine
>C=N	includes Imine
>N−C=O	includes Amide
−NO & $-NO_2$	includes Nitroso, Nitro
−SH	includes Thiol
−S−	includes Sulfide
>C=S	includes Thiocarbonyl
$=CH_2$	includes Methylene
>C=C<	includes Ethylene
	includes Aromatic
CH only	includes Hydrocarbon

Card deck C provides for more precise searching by the presence of peaks. A search is made by selecting those cards that represent the characteristics of an unknown curve and superimposing these over a light source. For example, one might look for standard compounds having an hydroxyl group and exhibiting peaks at 14.1, 12.3, and 8.2 μ in their spectral curves. By superimposing cards representing these characteristics one can identify those compounds that would satisfy the search question.

2. Superimposed Coding for Edge-Notched Punched Cards

Superimposed coding is a concept that involves storing the codes related to a document in the same place, i.e., codes are superimposed in the same field on top of each other. For example, assume codes that are based on a combination of two numbers in the 1 to 20 ranges. If a document were indexed by the descriptors "chemistry" and "theory" and the code for chemistry was 4-16 and the code for theory was 4-20, the codes for these descriptors would overlap or be superimposed on top of each other. A representation of this on an edge-notched punch card might appear as follows:

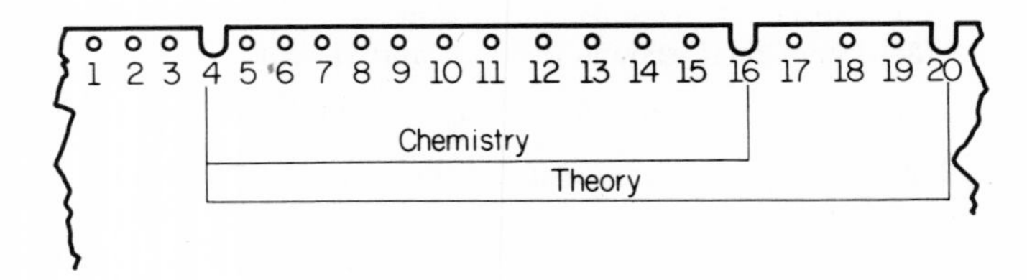

Fig. 1. Superimposed coding.

In an edge-notched card system, each document surrogate (a representation of a document such as a document number, a bibliographic citation, or abstract) is represented on a card and all subject terms (or codes for the terms) relative to that document are indicated on that card. (This is the reverse of coordinate indexing.) A superimposed coding system employs relatively few subject terms or descriptors. This technique is particularly suitable where coding space is limited as in the case of edge-notched punch cards. Two examples of information services that employ superimposed coding on edge-notched punch cards are the Gas Chromatography Abstracting Service of Preston Technical Abstracts Inc., and the RINGDOC system of Derwent Publications, Ltd.

3. Link and Role Indexing

Link and role indexing was developed to overcome problems of false drops resulting from false coordination of terms. A link is a code that is used to provide syntax to a group of index terms. It ties together terms that are related in the same logical phrase or context. Role indicators are codes that tell what role an index term plays. As an example of the use of links and roles, consider document X which deals with: "Research on the effect of heat on the degradation of polymers, and the effect of radiation on the degradation of styrofoam." Assume the following role definitions:

2 — independent variable studies for its effect on
8 — research on . . . , investigation of . . . ,
9 — dependent variable studied for how it is affected.

The indexing for document X would be as follows:

Link	Role	Index term
A	8	degradation
A	2	heat
A	9	polymers
B	8	degradation
B	2	radiation
B	9	styrofoam

Link and role indexing requires that the document analyst be very familiar with the subject field. It is time-consuming but if done properly can permit very refined searches. Link and role indexing was used by the American Society for Metals (ASM) computer-based retrieval system.

C. Vocabulary Control and Nomenclature

To maintain consistency and to achieve an effective system, index terms must be controlled. There are two basic approaches to the establishment of vocabularies or index term lists for information systems. One involves the use of a prescribed list of index terms; the other involves the use of an open-ended though controlled vocabulary in which the vocabulary is generated as a result of indexing actual documents. There are examples of successful systems of both types; however, prescribed inflexible vocabularies or closed systems are more likely to be of use in subject fields where little change takes place. The lack of flexibility and use of a rigid predetermined hierarchical structure of the Dewey Decimal Classification system, for example, has greatly limited its utility for scientific collections. Because of today's expanding technology, open-ended vocabularies are more suitable for scientific and technical information systems.

When a new information retrieval system is started, only a representative sampling of the total collection is indexed initially. The resulting list of terms is then analyzed. The purpose of the analysis is to control the vocabulary, that is, eliminate redundancies, combine synonyms, introduce higher generic terms if necessary, break down generic terms into specific terms, determine the use of singular or plural form of words, or determine the accepted permutations and use of multiword terms, abbreviations, etc. The list provides the basis for a thesaurus, authority list, or controlled vocabulary.

The authority list is used to tell the indexer and the user which terms are approved and accepted as index terms. An authority list includes cross references to send the indexer or user from nonapproved synonyms or semantically related terms to the term that is used or accepted by the system. For example, if one looks in the Index Guide (which explains CA indexing policy and indicates which index terms are used in CA indexes) for magnetic resonance absorption he will find the entries "Magnetic resonance absorption by atomic nuclei" and "Magnetic resonance absorption by electrons." These are not accepted terms so he is sent by means of "see" references to the accepted term(s). The see reference under "Magnetic resonance absorption by atomic nuclei" is "see: Nuclear magnetic resonance." He is also told by means of "see also" references that additional related information may be found by checking several related terms: "see also: Knightshift; Magnetic relaxation; Microwaves, absorption; Nuclear quadrupole resonance; and Radiowaves, absorption."

Without a controlled vocabulary one might have to look for the information on electron-spin resonance under any one of the terms that make up the multiword index term or its synonyms. The lack of vocabulary control is one of the difficulties encountered in using KWIC (Keyword-In-Context) indexes which are computer generated indexes prepared automatically from titles. (The publication Chemical Titles includes a KWIC index produced by Chemical Abstracts Service.) Various authors use different synonymous terms for the same concept. A user, then, must search under all possible synonyms if he wants to conduct a thorough search of a KWIC index. The burden is put on the user to accommodate the variation in terminology used by authors in writing titles. Studies have been made which indicate that systems that operate from a controlled vocabulary obtain the highest relevance rates.

Subject terms are not the only keys used for conducting searches. Often searches are conducted using author names, corporate authors, journal identification, document number, date of publication, etc., as principal keys. Such searches are easy to conduct and do not require detailed content analysis or each document. However, as in the case of subject index terms, consistency is necessary and there has to be a unified system for generating and storing bibliographic citations. One must, for example, adhere to one system for abbreviating journal names and corporate author names. These in effect become authority lists or controls for abbreviations.

Some standards for abbreviation have come into wide use and are accepted in many organizations, e.g., the Chemical Abstracts Service Source Index and the American Society of Testing and Materials CODEN list which provides unique five-character codes for journal names.

Control must be exercised in the assignment of all identifiers used in retrieval systems with respect to the abbreviations adopted, numbering systems for dates, pages and document identification, punctuation, capitalization and permuted order or linear arrangement of any multicomponent terms (e.g., Jones, A vs A. Jones or *Chemical Abstracts* vs *Abstracts*, *Chemical*).

In the field of chemistry the use of a controlled vocabulary includes adoption of a *system of nomenclature*. One cannot interchange preferred chemical names, common names, and trade names. If one system is not adhered to, a searcher looking for information on a specific compound would have to search not under one name but under all possible names. This becomes extremely difficult because trade names cannot be predicted. Ten different pharmaceutical companies may market the same drug under ten different names. The need for ground rules and the use of authority lists for indexing is obvious inasmuch as the lists greatly simplify the search procedure. The *Chemical Abstracts* (CA) compound name which is largely based on the IUPAC (International Union of Pure and Applied Chemistry) system is widely accepted as a standard and is used in both the published CA indexes and in their Compound Registry System.

The various indexing and nomenclature practices used in source documents is illustrated by a search for information relative to 3-chloropropene.

IUPAC Name	Molecular Formula	Structure
3-chloropropene	C_3H_5Cl	H, H, C=C, H, C, H, H, Cl

CA (Vol. 76) uses as its accepted entry "1-propene, 3-chloro-"; a CA cross reference one might use to find the accepted term is listed as "Allyl chloride (see 1-propene, 3-chloro-)." The *Handbook of Chemistry and Physics*, on the other hand, uses "Allyl chloride" as its accepted term and a "see" reference to it is listed as "Propene, 3-chloro- (see Allyl chloride)." The *Merck Index* also uses "Allyl chloride" as its accepted term and cross references to it are listed as "3-chloropropylene (see Allylchloride)", "chlorallylene (see Allylchloride)."

The *SOCMA Handbook* uses 3-chloropropene, which is cross indexed by Allylchloride, Chlorallylene, 3-Chloroprene, and 3-Chloropropylene. It also gives the Registry Number 107-05-1. Beilstein's *Handbuch der Organischen Chemie* lists 3-chloropropene-1 with the synonyms γ-chloropropylene and allyl chloride.

The need for controls and standards is even more important when information collections are processed by computers. Computers are machines; they cannot "read between the lines" or make inferences and associations. They conduct searches (match words) according to rigid programmed instructions.

D. User Aids

Records are often maintained regarding the frequency of occurrence of index terms. These frequency lists are useful to the indexers because they indicate the need for further breakdown of very frequently used terms into more specific terms. Term frequency lists, vocabulary lists with cross referencing, standard abbreviation lists as well as dictionaries, product indexes, and standard reference works are often referred to as user aids. They are useful for selecting search terms for both manual and machine searches. System vocabulary lists and abbreviation lists tell the user which identifiers are actually used in the system, and frequency lists provide the user with a rough index to the number of documents he may find using any specific term.

Following, or simultaneously with indexing, coding may take place. Codes are used in traditional classification systems and in many nonconventional indexing systems.

E. Coding for Retrieval

Coding is a substitution process — a surrogate representation is substituted for an identifiable entity. Coding is the application of a set of rules for converting data from one set of symbolic representations to another. In information systems codes may be used for subject terms, classes, file location, or any bibliographic or descriptive information. Coding is usually used to provide a shortened version of the original information, to provide uniformity and systematization in the presentation of information, and to facilitate the use of automatic equipment.

Machine systems for information retrieval involve not only the codes used for subject terms, descriptive information, link and role indicators, etc., but also the character sets or numeric representation of these for machine recognition. Character sets are groups of numbers, characters, letters, or symbols for which bit patterns have been assigned according to a set of rules. Machines recognize numeric codes such as binary coded decimal, EBCDIC (extended binary coded decimal interchange code), ASCII (American National Standards Institute Interchange Code), etc. The character set used is dependent on the specific machine.

Codes can indicate meaning, size, physical position or location in a file, position in a hierarchy, or position in time. Codes can be used to condense and categorize information; any type of desired qualification can be included in an encoding scheme. Substituted for words or sentences, codes permit a savings of space in a machine file because codes are generally shorter than that which they represent. They regularize the structure of the file when they are of uniform length. Problems of synonymy and ambiguity are eliminated from the retrieval system since they are resolved manually during the encoding procedure. If word control is exercised, however, reference to a word guide, thesaurus, or code directory is required.

F. Automatic Indexing and Abstracting

Because the intellectual process of indexing is time consuming, requires highly skilled personnel, and is, therefore, expensive, considerable effort has been expended in the area of automatic indexing.

Research efforts have led to the development of permuted title (Key Word in Context or KWIC) indexes, citation indexes, and automatic indexing from full text. The successes that have been achieved in automatic indexing are generally considered successes on the basis of time savings, not on the basis of indexing quality. Basically, these systems employ rules for the selection of key words, scan tape files to select the words, screen out noise words, and prepare and print an index list in either inverted or serial order.

A computer can "index" by selecting and printing out lists of terms taken from titles, abstracts, and text on the basis of: comparison with an exclusion-word dictionary of common function words; comparison with an inclusion-term dictionary of content words generally within a definable technical subject area; frequency of occurrence; infrequency of occurrence; occurrence in relation to physical proximity to other terms; co-occurence of word pairs; or combinations of the above.

A computer can only match or select terms that are provided as input text or as dictionary look-up terms. A human indexer can judiciously select and assign terms inclusive and beyond those that are actually used as input. A human indexer can make associations, inferences, and judgments that a computer cannot. A human can also select or exclude index terms based on a variety and a combination of semantic, syntactical, and contextual connotations and considerations.

A KWIC index or permuted title index is prepared by sorting each of the principal words in a title and printing out a listing of principal words in alphabetical order, with the rest of the title permuted around the principa- word (or a specified number of characters in the title). KWIC indexes are limited by the inherent limitations of titles, by the fact that few index words are involved, and by the lack of vocabulary control. They are, however, inexpensively and quickly produced; they display words in context, and "canned" programs exist for many computers. KWIC indexes are often used for current awareness operations.

A citation index is a bibliographic listing, usually arranged by author, that provides citations to the reference. It traces the history of an article by noting other articles that have cited it.

Autoindexing from full text involves the automatic selection from text of words and/or word pairs for use as index terms. When this is done, however, human editing is required to remove illegitimately paired words, to add the third and/or fourth words to multiword terms that required more than paired terms, and to prepare all "see" and "see also" cross references. "No program has yet been devised to provide program-directed automatic indexes or abstracts to an operating information system directly from full text documents." [1]

Words, and words in phrases, sentences, and paragraphs, are counted and compared by a computer in a physical manner rather than as a human would. Computers can manipulate and calculate data, tabulate and reproduce citations and documents, and store information. A computer can do nothing more than compare and count, unless additional coding and structure beyond natural syntax are supplied.

Since index terms are the principal means by which one gains access to items in a store of information, the importance of indexing regardless of what method is employed cannot be overemphasized.

G. Computerized Retrieval

The usefulness of computerized retrieval systems to chemists is to a large extent dependent on the adequacy and availability of data bases as well as the search systems for handling them. Improvements in computer technology and associated economic factors have led to the increased use of computers in the printing and publishing industry. Organizations that create machine-readable information tapes as part of their processing or publishing activity create data bases as a by-product. Some fifty major scientific and technical data bases are available at present, and many more will come into existence in the near future.

Data bases by themselves are useless. There must also be search systems for handling the data bases and providing desired output to users. The data bases must contain information that is useful (see Section III of this chapter for data base information); search systems must be capable of responding to user needs. A wide variety of search services and data bases have been made available to the public in recent years, so that the individual scientist can now identify the data base(s) he wants to search. If his own organization does not provide the service he can shop around for a center that handles that data base and provides the type of service and output he wants.

There are many advantages to using computerized retrieval systems and the services of information centers. Although the prospects for doing true content analysis, indexing, and abstracting by machine do not look encouraging, the usefulness of searching by machines has been well demonstrated. Techniques have been developed for overcoming probelms related to searching the uncontrolled terminology that is found on many data bases. These are described later in Section IV. Advantages of computer searching are given in Section V.

III. INFORMATION AND DATA

Among the sources of information a scientist uses is the "invisible college." This college is compromised of individuals who through personal acquaintance, shared experiences, and similar work activities, communicate with each other in a personal and informal manner. Private communication and private exchange characterize the functioning of the "invisible college." No treatment is given here to the "invisible college" and other private sources of information since there is no systematic way to approach such information. Our concern is with the publicly available sources and to a large extent with the publicly available machine-readable sources.

Also, since there are a number of good texts that deal with use of the traditional published literature, that aspect of chemical literature is treated very lightly here. Further information on the use of chemical reference material can be found in the publications of Bottle [2], Burman [3], Crane [4], Mellon [5], Walford [6], and Winchell [7].

A. Bibliographic/Document Related Data

1. Machine-Readable Bibliographic Data Bases for Chemists

There are many bibliographic document related data bases that are of potential use to chemists. Naturally, CA Condensates is probably the most obvious choice by the majority of chemists because it contains citations for all of the more than 300,000 abstracts published by CAS per year. Some data bases designed specifically for chemists are CA Condensates, CA Subject Index Alert (CASIA), Chemical and Biological Activities (CBAC), Chemical Titles (CT), Polymer Science and Technology (POST), Patent Concordance, CA Integrated Subject File (CAISF), Chemical Abstracts Service Source Index (CASSI), Index Chemicus Registry System (ICRS), and IFI Tapes. Descriptions of these files and a listing by file name together with addresses for these data bases as well as for data bases that contain some chemical or chemically related information are given below. Because chemistry interfaces with so many other sciences and technologies and because of the interdisciplinary character of research in the 1970s, the number of data bases of use to chemists is more extensive than one might anticipate.

These and other data bases are available for lease or license by individual companies, universities, or governmental bodies. Where the need is not sufficient to warrant operation of a search facility internally, data base search services can be purchased either directly from data base suppliers or from a variety of information centers. Further information regarding data bases can be obtained from the Survey of Scientific-Technical Tape Services by Carroll [8], from An Inventory of Some English Language Secondary Information Services in Science and Technology compiled by Lee [9], from the ASIS SIG/SDI Survey of Data Bases, or directly from the data base suppliers. Information regarding data base search services available from centers can be found in the ASIDIC Survey of Information Center Services by Williams and Stewart [10].

a. Data Bases Designed for Chemists.

CA Condensates
Chemical Abstracts Service
The Ohio State University
Columbus, Ohio

CA Condensates provides the following searchable information from the corresponding issues of Chemical Abstracts: titles of papers, patents, and reports; names and affiliations of authors and assignees; bibliographic citations; CA section numbers and entries from the CA Issue Keyword Subject Index. The service is designed to allow the user to scan rapidly the bibliographic citations and keyword content related to approximately one-third of a million new articles and patents each year.

CBAC (Chemical-Biological Activities)
Chemical Abstracts Service

CBAC is a file of abstracts that reports the current scientific and technical literature related to the field of biochemistry as covered by CA. The literature covered includes journals, patents, reports, and conference and symposium proceedings that report the interactions of chemical substances with biological systems in vivo and in vitro. Searchable information includes names and affiliations of authors and assignees, bibliographic citations, document titles and full abstracts, molecular formulas, names of substances, CA Section Numbers, and CAS Registry Numbers.

CT (Chemical Titles)
Chemical Abstracts Service

CT is the service that reports the titles of selected papers recently published in current chemical journals and have been selected to be abstracted in CA. CT is designed to enable chemists and chemical engineers to stay abreast of the latest reported advances in theory and technology by alerting them to current information appearing in about 700 of the world's respected chemically oriented journals. Searchable information includes document titles, names of authors, and bibliographic citations.

Chemical Abstracts Service Source Index
Chemical Abstracts Service

The computer-readable form of the Chemical Abstracts Service Source Index identifies the original source literature of chemistry and chemical engineering. The service is designed to help users locate the libraries that hold the journals, conference proceedings, and monographs that constitute the source literature. Searchable information includes, but is not limited to, bibliographic and library holdings information, present and former publication titles abbreviated according to American National Standards Institute rules, CODEN, holdings of contributing libraries, publisher names and addresses, publications titles according to American Library Association cataloging rules, and language of publication.

IFI Tapes
IFI/Plenum Data Corporation
1000 Connecticut Avenue, N. W.
Washington, D. C.

The IFI Tapes contain information on U. S. Chemical and chemically related patents with foreign equivalents. Searchable data elements include descriptions, assignees, patent numbers, and U. S. Patent Office class codes. The tape records include patent titles and some Chemical Abstract reference numbers.

ICRS (Index Chemicus Registry System)
Institute for Scientific Information
325 Chestnut Street
Philadelphia, Pennsylvania

ICRS contains chemical information from 2500 journals. It reports new compounds, new reactions, and new methodologies in chemistry. Searchable data elements include analytical data codes, compound uses, author affiliations, journal titles enriched with added descriptions, Wiswesser Line Notations (WLN), and molecular formulas.

POST (Polymer Science and Technology)
Chemical Abstracts Service

POST contains abstracts that report the current scientific and technical literature related to the field of macromolecular chemistry. Coverage includes the worldwide journal and report literature, plus patents from 26 countries that describe research, development, production, uses, equipment, and all other aspects of macromolecular chemistry. Searchable information includes names and affiliations of authors, bibliographic citations, document titles and full abstracts, molecular formulas, names of substances, CA Section Numbers, and CAS Registry Numbers.

Patent Concordance
Chemical Abstracts Service

The *Patent Concordance for Chemical Abstracts* starting with Volume 56 (January-June 1962) is available in computer-readable form. A Patent Concordance refers all subsequent patents (by number and issuing country) to the abstract of the first equivalent patent published previously in *Chemical Abstracts*.

CAISF (CA Integrated Subject File)
Chemical Abstracts Service

CAISF is the computer-readable counterpart to the printed indexes to *Chemical Abstracts*. Designed for the searching of Volume Index entries to CA abstracts, the CAISF provides the information contained in the corresponding printed volume-level Chemical Substance and General Subject indexes, plus molecular formulas and CA Section Numbers. Searchable information includes chemical and nonchemical concepts, CA Index Names, molecular formulas CAS Registry Numbers, and CA Section Numbers. Each volume of the CAISF refers to a specific volume of CA and carries the same numeric designation (e.g., CAISF-76 refers to CA Volume 76). Each volume of the CAISF is divided into two parts. Part A contains all chemical substance index entries; Part B, all general subject index entries. Sort Keys are provided in each Part to allow the merging of two or more volumes of the CAISF.

CASIA (CA Subject Index Alert)
Chemical Abstracts Service

The CA Subject Index Alert (CASIA) is a computer-readable, current-awareness file issued biweekly that contains Volume Index entries to Chemical Abstracts (CA). These entries are arranged document by document in CA abstract number order. CASIA has no corresponding printed counterpart, although all entries included in CASIA issues will appear later (no longer in document order but in Index order) within the printed Volume Indexes to CA and in the equivalent computer-readable service, CA Integrated Subject File.

Entries in CASIA are grouped in sets, each set containing these Volume Index entries corresponding to an abstract published in the printed CA: the CA Index Names, molecular formulas, CAS Registry Numbers, and the General Subject Index entries. All index entries for a given paper or patent appear at one point on the tape associated with the CA abstract number that refers to the document. These sets appear in ascending CA abstract number sequence and although in consecutive order, the sets are frequently non-contiguous. Each issue of CASIA contains those sets of Volume Index entries that have been completed and edited since the previous CASIA issue. Since the time required to complete and edit Volume Index entries to a particular abstract varies, there is no time correspondence between an abstract number in CASIA and the same abstract number in the printed CA issues. Normally, a CASIA issue will contain abstract numbers from several CA issues, and possibly more than one CA volume period.

b. Data Bases of Use to Chemists.

Abstract Bulletin of the Institute of Paper Chemistry
(monthly)
The Institute of Paper Chemistry
1043 East South River
Appleton, Wisconsin 54911

Abstract Bulletin of the Institute of Paper Chemistry, Keyword Supplement
(monthly)
The Institute of Paper Chemistry
1043 East South River
Appleton, Wisconsin 54911

Air Pollution Abstracts
(biweekly)
Environmental Protection Agency
Washington, D. C.

<u>API Abstracts of Refining Literature</u>
(monthly)
American Petroleum Institute
555 Madison Avenue
New York, New York 10022

<u>API Abstracts of Refining Patents</u>
(monthly)
American Petroleum Institute
555 Madison Avenue
New York, New York 10022

<u>ASCA (Automatic Subject Citation Alert)</u>
(weekly)
Institute for Scientific Information
325 Chestnut Street
Philadelphia, Pennsylvania 19106

<u>BA Previews</u>
(Biological Abstracts: semimonthly)
(BioResearch Index: monthly)
Biosciences Information Service
2100 Arch Street
Philadelphia, Pennsylvania 19103

<u>Bibliography of North American Geology</u>
U. S. Geological Survey
Department of Interior
Washington, D. C.

<u>CA Condensates</u>
(weekly)
Chemical Abstracts Sercice
The Ohio State University
Columbus, Ohio 43210

<u>CAISF (CA Integrated Subject File)</u>
(semiannually)
Chemical Abstracts Service
The Ohio State University
Columbus, Ohio 43210

<u>CAIN (Cataloging and Indexing System)</u>
(monthly)
National Agricultural Library
Food and Nutrition Information Center
Beltsville,
Maryland 20705

<u>CBAC (Chemical-Biological Activities)</u>
(biweekly)
Chemical Abstracts Service
The Ohio State University
Columbus, Ohio 43210

<u>Central Patents Index — Minitapes</u>
(weekly)
Derwent Publications, Ltd.
Rochdale House
128 Theobalds Road
London WC1X 8RP
England

<u>Chemical Abstracts Service Source Index</u>
Chemical Abstracts Service
The Ohio State University
Columbus, Ohio 43210

<u>CT (Chemical Titles)</u>
(biweekly)
Chemical Abstracts Service
The Ohio State University
Columbus, Ohio 43210

<u>CITE (Current Information Tapes for Engineers)</u>
(monthly)
Engineering Index. Inc.
345 East 47th Street
New York, New York 10017

<u>COMPENDEX (Computerized Engineering Index)</u>
(monthly)
Engineering Index, Inc.
345 East 47th Street
New York, New York 10017

<u>DDC Tapes</u>
(semimonthly — limited availability)
Defense Documentation Center
Arlington, Virginia

EXERPTA MEDICA
(weekly)
3i Company
2101 Walnut Street
Philadelphia, Pennsylvania 19103

FSTA (Food Science and Technology Abstracts)
(monthly)
International Food Information Service
6 Frankfurt
Main-Niederrad
Herriotstrasse
West Germany

Geophysical Abstracts
U. S. Geological Survey
Department of Interior
Washington, D. C.

GEO•REF (Geological Reference File)
(monthly)
American Geological Institute
2201 M Street, N.W.
Washington, D. C. 20037

GRA (Government Reports Announcements)
(semimonthly)
National Technical Information Service
U. S. Department of Commerce
Washington, D. C. 20230

ICRS (Index Chemicus Registry System)
(monthly)
Institute for Scientific Information
325 Chestnut Street
Philadelphia, Pennsylvania 19106

IFI Tapes (Information For Industry)
(bimonthly)
IFI/Plenum Data Corporation
1000 Connecticut Avenue, N. W.
Washington, D. C. 20036

INIS (International Nuclear Information System
(monthly)
International Atomic Energy Agency
P. O. Box 590
A-1011 Vienna
Austria

ISI Science Citation Index Source and Citation Tapes
(weekly)
Institute for Scientific Information
325 Chestnut Street
Philadelphia, Pennsylvania 19106

INSPEC
(semimonthly)
The Institution of Electrical Engineers
Savoy Place
London, WC2R 0BL
England

MARC (Machine Readable Cataloging) Distribution Service
(weekly)
Library of Congress
Washington, D.C. 20540

MEDLARS (Medical Literature Analysis and Retrieval System)
(monthly)
National Library of Medicine
Bethesda, Maryland 20014

METADEX (Metals Abstracts Index)
(monthly)
American Society for Metals
Metals Park, Ohio 44073

NASA Tapes
(semimonthly — limited availability)
National Aeronautics and Space Administration
Washington, D.C. 20546

Nuclear Science Abstracts
(semimonthly)
Atomic Energy Commission
Division of Technical Information
Washington D.C. 20545

PANDEX
(weekly)
Macmillan Information
866 Third Avenue
New York, New York 10022

Patent Concordance
(periodically)
Chemical Abstracts Service
The Ohio State University
Columbus, Ohio 43210

PESTDOC
(monthly — available on punched cards for conversion to tape)
Derwent Publications, Ltd.
Rochdale House
128 Theobalds Road
London WC1X 8RP
England

POST (Polymer Science and Technology)
(biweekly)
Chemical Abstracts Service
The Ohio State University
Columbus, Ohio 43210

RINGDOC
(monthly — available on punched cards for conversion to tape)
Derwent Publications Ltd.
Rochdale House
128 Theobalds Road
London WC1X 8RP
England

Selected Water Resources Abstracts
(biweekly)
Environment Protection Agency
Washington, D. C.

SPIN (Searchable Physics Information Notices)
(monthly)
National Information System for Physics & Astronomy
American Institute of Physics
335 East 45th Street
New York, New York 10017

Textile Technology Digest
(quarterly)
Institute of Textile Technology
Charlottesville, Virginia

VETDOC
(monthly — available on punched cards for conversion to tape)
Derwent Publications, Ltd.
Rochdale House
128 Theobalds Road
London WC1X 8RP
England

World Textile Abstracts
(semimonthly)
Shirley Institute
Didsbury
Manchester M20 8RX
England

B. Structural Data

1. Representing Chemical Structures for Storage and Retrieval

The chemical structure is the least common denominator of chemical information; in searching the chemical literature it is most often the key that is used. "Although there are many other aspects of chemistry, the chemist is usually concerned with the molecular identity of materials, and it is this kind of subject matter about which he most frequently communicates." [11] Data about chemicals —properties, effects, processes, etc. —are meaningful when associated with specific compounds or classes of compounds and their structures.

Much of the time and money that has been spent on information processing in chemistry has gone toward the development of systems for communicating structural information of chemical compounds. Structures are the natural language of chemists and the best way of communicating them is by means of graphic representations. While structures can be drawn unambiguously, naming, coding, and classifying them for storage and retrieval is not so straightforward. Considerable effort has been put into developing methods of representing, indexing, designating, and labeling structures in computer files in order to locate specific structures, structural components, or groups of similar structures. The major types of systems that have been worked on in recent years are fragmentation, linear notation, and topological coding systems.

Structure information can be communicated by means of graphical diagrams, nomenclature, fragmentation codes, or complete representations including notations or ciphers, connection tables, and topological codes. The structural diagram is a two-dimensional graphical diagram that identifies the elements present in a substance, indicates which atoms are connected to each other in a molecule of the substance, and shows the type of bonds connecting the atoms.

a. Graphical Diagrams. While the diagram or picture is good for communication of information, it is difficult to discuss a structure without the aid of paper and pencil, and it is difficult to arrange and store diagrams in an order that will be widely useful for searching.

b. Nomenclature. To assist in the process of communicating information about structures, systems for naming structures were developed. Systematic nomenclature is essential for oral communication, but it is beset by certain difficulties: a given nomenclature system requires a set of rules and the rules may take some time to learn; the rules will not necessarily be applied in exactly the same way by different chemists or indexers; new classes of compounds may come into being for which rules do not yet exist; and names do not provide a good basis for grouping or classifying compounds according to structural components.

c. Fragmentation. Fragmentation techniques and notation techniques were developed in order to identify or access compounds according to structural components via alphanumeric codes. "Fragmentation methods are based on the principle that various atoms and groupings of atoms occur in molecules and that co-occurrences of such groupings are important in retrieving structure." [12] Preselected fragments (the major ones) are recorded, but multiple occurrences of the same fragment are not, nor is context information; hence, the structure cannot be reconstituted from the fragments. One can locate a compound or group of compounds by coordinating two or more significant fragments (atoms or groups of atoms) that have been used in the system. Fragmentation systems are easy to learn, but the complete structure is not represented, i.e., not all fragments are used. Fragmentation techniques are used by the National Cancer Institute, Abbott Laboratories, Kodak Research Laboratories, and others.

d. Linear Notation. The complete structure can be represented by means of ciphers or linear notation systems such as the Wiswesser Line Notation (WLN) or the IUPAC (International Union of Pure and Applied Chemistry) Dyson System. Ciphers or linear notations are one-dimensional arrays of symbols that can unambiguously represent the topology of molecules. Such systems take time to learn and occasionally rules have to be changed to accommodate new circumstances. A given notation is presented in a specific linear order according to the rules; in a manual file it would be alphabetized according to the first letter. In order to locate other components of the compound it would be necessary to have multiple cross indexes or to prepare computer-generated permuted notation indexes.

e. Topological Codes. Another way of representing complete structures is by use of topological codes where atom-to-atom connections are described with atoms as nodes and connections or bonds representing branches in a network. This topological representation is a two-dimensional array of characters presenting atom-by-atom and bond-by-bond the parts of a molecule either in list form or as a set of matrices. The atomic groups and connection information make up a connection table. A variety of topological systems have been developed at Chemical Abstracts Service, du Pont, and other organizations.

Explanations of specific systems employing fragmentation, linear notation, and topological codes together with an excellent bibliography are presented in Chemical Structures Information Handling, published by the National Academy of Sciences, Washington, D. C., 1969 [13].

Examples of different representations of a chemical structure were presented by Holm [14] in a paper presented at the Conference on Computers in Chemical Education and Research at Northern Illinois University, 1971.

Graphical Diagram:

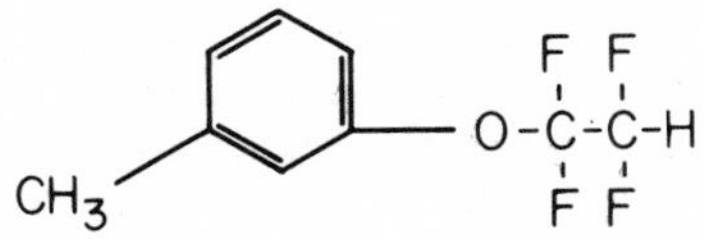

F F
O-C-C-H
F F
CH_3

Nomenclature: Phenetole, α,α,β,β,
Tetrafluoro-m-Methyl

Fragmentation: Ether
Fluorine
Benzene Ring 292
1-1 Configuration
1-2 Configuration

Notation: FUFXFFOR C

Topological Code:

C	1	1
O	1	1
F	1	1
F	1	1
F	2	1
F	2	1
C	3	1
C	8	L
C	8	L
C	9	L
C	10	L
C	11	L
C	11	L
	12-13	L

Fig. 2. Chemical structure representations.

2. Structure Files for Computer Searching

All of these representations can be put in machine-readable form and manipulated with varying degrees of expense and success. A computer-based system for handling information about chemical structures which will have national impact on chemists is the CAS Registry System. As of late 1973, CAS has assigned CAS Registry Numbers (unique, machine-assigned numbers) to some two and one-half million compounds of a possible three to four million compounds. These records include names, molecular formulas, structural information, and CAS Registry Numbers which are used to link all substance related information items in the file. Files such as CAISF contain links between CA abstract numbers and CAS Registry numbers. The file could be searched on the basis of name, structure, fragments, structural bond requirements (e.g., ring must be aromatic, etc.), structural properties (e.g., functional groups, number of carbon atoms, etc.) heteroatoms or Registry Numbers.

The use of Registry Numbers is going to add a new dimension to virtually all CAS product publications and machine-readable data bases as well as many others. This will simplify some of the problems one encounters when searching a data base via ambiguous nomenclature. Further work with the Registry and Substructure Search System has been done by the Department of the Army, the University of Georgia, and the National Cancer Institute, where an interactive on-line system has been developed which permits the user at the terminal to draw chemical fragments, impose restrictions on them, and interactively search the file to find structures containing the required fragments [15].

A more sophisticated system for working with structures on a terminal is being studied by Brookhaven National Laboratories and Texas A and M University on a National Science Foundation grant. They propose making three-dimensional views of structures available to users in small institutions by means of a computer library hooked up to a standard telephone network.

C. Numeric and Analog Data

Students of chemistry are familiar with the type of information that is recorded in the form of digitized numeric data. They are interested in properties such as melting points, bpiling points, and molecular weights; they are concerned with analog data obtained from infrared, nuclear magnetic resonance, mass spectrometry, gas chromatography, etc. Once analog data has been reduced or digitized, it then can be handled in an information storage and retrieval system.

Machine handling of numeric data is an easier task than handling bibliographic/document-related data or structural data. This is true because computers are number-oriented — there is no loss of meaning when converting from the representation of numbers on paper to the representation of numbers in a machine system. The communication problem is much greater when converting natural language words that have specific meaning in a specific context, or when converting structural diagrams to a machine representation.

A wide variety of useful data compilation and information systems exist for retrieving chemical data. Many of these printed versions are already familiar to chemistry students, such as the Handbook of Chemistry and Physics published by the Chemical Rubber Company. Following are lists of Handbooks and Compilations and Analytical Data Sources of which the chemistry students should be aware. The use of such resources in hard copy form is usually straightforward and does not warrant further explanation here. There are, however, a number of standard printed handbooks that are currently or plan to convert their data to machine-readable form. The Chemical Rubber Company, for example, is converting a portion of the handbook contents to magnetic tape; search services from such tapes will be abailable in the not-too-distant future. As the cost of computer time decreases, storage capabilities increase, and new and cheaper ways of inputing data (optical character readers and direct input devices) are developed, more and more publishers will join the ranks of the data base producers. Handbooks, compilations, and data resources will be in computer-readable form, and the possibilities for manipulating the data, making new correlations, etc., will be tremendously increased and speeded up. The potential will be tremendous.

1. Analytical Data Sources

Preston NMR Abstract Service
[Edge-notched cards (double row), tape]
World literature on all aspects of NMR
Preston Technical Abstracts, Inc.
909 Pitner
Evanston, Illinois 60202

DMS NMR Literature Service
(Printed literature lists with optical coincidence index)
Comprehensive coverage of literature of NMR, EPR and
Nuclear Quadruapole Resonance
Documentation of Molecular Spectroscopy
British Editor
Butterworths
88 Kingsway
London W.C. 2
England

Preston Gas Chromatography Abstract Service
(Edge-notched cards, microfilm, tape)
Comprehensive coded abstracts on gas chromatography
Preston Technical Abstracts, Inc.
909 Pitner
Evanston, Illinois 60202

DMS Current Literature Service in IR, Raman & Microwave
(Printed literature lists with optical coincidence indexes)
Comprehensive coverage of literature of IR, raman and microwave spectroscopy, documentation of molecular spectroscopy
British Editor
Butterworths
88 Kingsway
London W. C. 2
England

Infrared and Ultraviolet Spectral Index Cards
American Society for Testing and Materials
Philadelphia, Pennsylvania

Sadtler Infra-Red Spectra
(Printed and tape)
Infra-red Spectra Collection
Sadtler Research Laboratories, Inc.
3316 Spring Garden Street
Philadelphia, Pennsylvania 19104

DMS (Documentation of Molecular Spectroscopy) Service
(Double edge-notched cards)
Comprehensive collection of infra-red spectra;
documentation of molecular spectroscopy
Butterworths
88 Kingsway
London W. C. 2
England

Sadtler NMR Spectra
(Printed and tape)
NMR spectra
Sadtler Research Laboratories, Inc.
3316 Spring Garden Street
Philadelphia, Pennsylvania 19104

Sadtler Ultraviolet Spectra
(Printed)
Reference UV spectra
Sadtler Research Laboratories, Inc.
3316 Spring Garden Street
Philadelphia, Pennsylvania 19104

<u>Mass Spectrometry Bulletin</u>
(Printed and tape)
Literature of mass spectrometry and allied topics
Mass Spectrometry Data Centre
AWRE
Aldermaston
Berkshire
England

2. Handbooks and Compilation

<u>Analiticheskaya Khimiya Elementov</u>
Acad. Sci. U.S.S.R.

<u>Beilstein's Handbuch der Organischen Chemie</u>, 4th ed.
Beilstein Institut, Germany

<u>Perry's Chemical Engineer's Handbook</u>
McGraw-Hill Book Company, New York, New York

<u>Colour Index</u>
Society of Dyers and Colorists, Bradford, England

<u>Comprehensive Biochemistry</u> (Florkin and Stotz)
Elsevier Publishing Company, Amsterdam, London, New York

<u>Comprehensive Treatise on Inorganic and Theoretical Chemistry</u>
(J. W. Mellor) Longmans, Green & Co., Inc.

<u>Dictionary of Chemical Engineering</u>
Japan

<u>Dictionary of Organic Compounds</u>
Pollock and Stevens, Oxford Univ. Press

<u>Handbook of Chemistry and Physics</u>
Chemical Rubber Company, Cleveland, Ohio

<u>Handbuch der Analytischen Chemie</u> (Fresenius & Jander)
Springer-Verlag, OHG, Berlin, Germany

<u>Inorganic Syntheses</u>
John Wiley and Sons, Inc., New York

<u>International Critical Tables of Numerical Data, Physics, Chemistry and Technology</u> (E. W. Washburn)
McGraw-Hill Book Company, New York

<u>Macromolecular Syntheses</u>
John Wiley & Sons, Inc., New York

The Merck Index of Chemicals and Drugs
Merck and Co.

Methoden der Organischen Chemie (Houben Weyl)
Georg Thieme Verlag, Stuttgart, Germany

New Drugs in Japan
Pharmaceutical Daily News, Japan

Numerical Data and Functional Relationships in Science and Technology (Landold-Bornstein)
Springer Verlag OHG, Berlin, Germany

Organic Reactions
John Wiley & Sons, Inc., New York

Organic Syntheses
John Wiley & Sons, Inc., New York

Registry Handbook
Chemical Abstracts Service, Columbus, Ohio

Ring Index
American Chemical Society

Selected Values of Physical and Thermodynamic Properties of Hydrocarbons and Related Compounds (F. D. Eossini et al.)
Carnegie Press, Carnegie Institute of Technology, Pittsburgh, Pa.

SOCMA Handbook
American Chemical Society

Spravochnik Khimika (Chemical Handbook)
U.S.S.R.

Standard Methods of Chemical Analysis (Furman and Welcher)
D. Van Nostrand Co., Inc., Princeton, New Jersey

Standard Spectra
Sadtler Research Laboratories, Philadelphia, Pa.

Synthetic Methods of Organic Chemistry (W. Theilheimer)
Interscience Publishers, Inc., New York

Tables annuelles de constants et donée numerique,
Tables de constants et donées numeriques constants selectionnées
C. Marie, Gruthier-Villars and Cie, France

Technique of Organic Chemistry (Weissberger)
Interscience Publishers, Inc., New York

Thermophysical Properties of High Temperature Solid Materials (Y. S. Touloukian)
Macmillan Company, New York

Traite de Chemie organique
V. Grignard et al. (eds.), Masson and Cie., Paris, France

Treatise on Analytical Chemistry (Kolthoff & Elving)
Interscience Publishers, Inc., New York

IV. PREPARING A QUESTION FOR COMPUTER SEARCHING

The previous section identified a number of computer-readable data bases available to chemists. If a chemist wants to use a data base he must formulate his search question in a manner that is amenable to the system.

When preparing a question for a computerized retrieval system the person preparing the search question should be aware of both the searchable contents of the data base as well as the specific features of the search program in order to make optimal use of both. The user question, prepared for machine searching, is often referred to as an interest profile since it represents a profile of the user's area of interest. The term "profile" is used here to represent a search question, whether it is to be processed for current awareness or for a retrospective (historical) search. A computerized current awareness system is usually referred to as SDI (selective dissemination of information).

A profile is the input submitted to the search system and the output of the search is usually bibliographic citations that match the terms and logic of the profile printed on paper or cards or displayed on a terminal. Output may be as minimal as citation numbers which direct the user to another file to obtain further information, or it may consist of citations, citations with index terms, abstracts, or full text.

A search profile is a set of logically associated search terms that describe the user's interests. A profile provides the detailed specifications for retrieving items from a data file. A search term can be a word, word fragment, phrase, author, journal, or any other data element appearing in a title, abstract, text, or keyword listing.

The terms are associated in a logic expression which specifies conjunction (AND) and disjunction (OR) between pairs of terms. A NOT operator can preclude the appearance of a term on data elements. The logic equation must be satisfied for a citation to be retrieved and disseminated to a user.

A computer program compares the terms in the profile with the items in a machine-readable file and retrieves those items that meet the selection criteria of the user. The profile is a key that unlocks the store of information in the file.

A profile can be general or specific, depending on end-use. A general profile is defined by broad subject areas or subsets of these areas. A general profile may be processed in order to obtain output that would satisfy a group of users whose interests are in the same general area. Such profiles are often referred to as group profiles; since copies of the output are distributed to multiple users the service can be sold at a lower unit price than the service for an individualized or custom search. An SDI service using more than 150 subareas of engineering is offered by Engineering Index and sold under the name Card-a-Lert. The United Kingdom Chemical Information Service (UKCIS) of the University of Nottingham produces macro-profiles on electron spin resonance, steroids, photochemistry, and several other subspecialties of chemistry. NASA produces group profiles called SCANS (Selected Current Aerospace Notes) on 189 topics in aerospace. Other organizations that produce group or standard profiles are the National Technical Information Service (NTIS), the Science Information Exchange, and the Institution of Electrical Engineers in England.

Inasmuch as the cost of machine searching is relatively high, many more group profile services will probably be developed in the near future for mass consumption. The content of group profiles may change as data bases are merged and combined; for example, a subspecialty group in ecology might require citations from CA Condensates, Engineering Index's Compendex, Biological Abstracts, BA Previews, and ISI's Source Tapes.

A. SDI and Retrospective Search

A current awareness system is a retrieval system that selects information from a periodically generated file on a scheduled basis in accordance with an interest profile prepared by or for a user. An automated current awareness system is called an SDI (selective dissemination of information) system. SDI is one of the most powerful of the recently developed tools for providing an individual with desired information from the growing literature resources.

A retrospective search system is one that selects information from a given file in accordance with a profile on a one-time basis. When asking a question for retrospective search the user should determine the time period he wants covered in addition to all of the other search parameters he would define for an SDI search. A retrospective search is often a good way of preparing for SDI. Once the user gets all of the background citation in his area of interest from a retrosearch, he can examine the output and determine whether to narrow or broaden his profile, and then use it for keeping himself up to date via an SDI service from the same data base.

B. Profile Development

The process of developing an effective profile is an empirical technique by which a user's interests are defined in accordance with the contents of a data file and search results are evaluated in accordance with the relevance of retrieved items. The semantic, syntactic, grammatical, and generic variations of the language of science are involved in both the user's profile and the data file. In addition, the language constraints of the data file, problems of classification and indexing, abbreviations and nomenclature add to the complexity of the search strategy. And in the case of an SDI service, these must be continuously monitored, because changes and additions that a data base supplier might make to the data base will have an impact on the effectiveness of a profile.

To facilitate the formulation of an effective profile, a five-step development process is suggested: (1) formulate question; (2) identify concepts and terms; (3) expand concepts and terms; (4) refine profile; (5) modify profile.

Steps one through five are needed for SDI, whereas only steps one through four are needed for retrosearch.

1. Formulate Question

A detailed but concise question is written to serve as a basis for the profile. The written statement in the language and terms familiar to the searcher expresses his particular interest. Putting his question in writing helps him to clarify his thoughts.

2. Identify Concepts and Terms

The question is examined to identify the concepts contained in the question. Concepts or ideas include terms relating to the subject matter of the question. These may involve elements, processes, materials, structures, functions, organisms, products, producers, agents, effects, reactions, compositions, and any other aspects required to express the "who, what, when, where, and why" of the question.

3. Expand Concepts and Terms

The concepts or terms identified in the previous stage are expanded to include synonyms, related terms, and narrower or broader terms that could be employed by various authors, abstractors, and indexers in referring to the same or similar concepts. Concept expansion is, in effect a translation of the question from the language of the user to the language of an information file. This step is greatly simplified when vocabulary controls have been exercised on the data base input.

Several sources are available as user aids to assist in concept expansion-translation. The user should make a careful study of his own information files and examine the titles, abstracts, and articles or documents that are known to be relevant to the questions in order to identify terms and phraseology used by authors in the field. He should also look at a few copies of the printed versions of the material to be searched; i.e., if Chemical Abstracts Condensates is the data base to be searched, he should examine several issues of the hard copy Chemical Abstracts. Most of the bibliographic data bases have hard copy counterparts. In fact, in many cases the machine-readable file is a by-product of creating the printed version. Checking the printed version will familiarize the user with editorial policies regarding coverage, terminology, file contents, and organization. After listing terms identified in his own files and the hardcopy version of the data base the user should then consult dictionaries, thesauri, glossaries, and word lists to find synonyms and alternate terminology that might be used to express his concepts. In addition to these, many data base suppliers and information processing centers provide search manuals and guides that are useful in preparing search terms for their data bases.

A reprint of Section IV from Chemical Abstracts Volume 76 Index Guide (Jan.-June 1972) entitled Naming and Indexing of Chemical Substances for Chemical Abstracts During the Ninth Collective Period (1972-1976) and the IUPAC 1957 Definitive Rules for Nomenclature of Organic Chemistry are useful tools for identifying chemical nomenclatures. Several other CAS and IUPAC nomenclature documents are available from CAS.

4. Refine Profile

The expanded concepts are reviewed for clarity, completeness, and compactness. Both abbreviated forms and expanded forms of terms are included as needed. Term truncation is employed where necessary and feasible. The terms are combined in a logical expression and depending on the search system, weights may be assigned to them. The resultant question is now ready for entry onto a coding form and subsequent submission to the search system.

5. Modify Profile

Profile construction is an art, and the achievement of acceptable results on the first attempt is often not possible. The evaluation of retrieved output can be made only by the user. The results of the evaluation can be used to reformulate the question and refine the keys to achieve a more effective search. The user should reformulate his question after each output until he is satisfied that the profile is producing the best possible output within the constraints of the data base content and the search system.

C. Profile-System Features

If one has selected a data base that contains information relevant to his area of interest, a good profile is the key to successful retrieval. Basically, a profile is made up of terms related to each other in a manner that expresses the user's question. Search terms may be of many types depending on the contents of the data base. Relationships between terms are usually expressed by means of logic.

The specific components of a profile will depend on the specific features or capabilities that have been programmed into a given search system. Features that are commonly found in retrieval systems are explained in the following. The specific manner in which profile coding is done will, of course, vary with the conventions used with a specific system.

1. Term Definition

Specific search programs may define a profile search term in any number of ways. It may be a single natural language word, an abbreviated word or acronym, a multiword term with or without hyphens, a phrase, a portion of a word, or there may be a restriction as to the total number of alphabetic and/or numeric characters that make up a term. Where a length limitation is imposed it is usually for purposes of economy, but the limit should not be one that will hinder effective retrieval. Some search systems limit search terms to a maximum of 20 characters, and this is reasonable as few terms exceed 20 characters. In CA, 94 percent of the terms fall within that range, and the average is only 10.8 characters.

Even if a user needed to use a term that exceeded the maximum he could accomodate the situation by breaking the term into two parts, use each part as a separate term and connect them together by means of AND logic. For example: 2,2-dimethyl 1, 3-propanediol could be represented as 2,2-dimethyl 1, 3-propanediol could be represented as 2,2-dimethyl 1,3-pro* plus *panediol (asterisk denotes truncation); however, in most instances the number of possible terms that might be retrieved by using only the first 20 characters of a 20+ character word would be unlikely to cause a problem.

Search terms in the profile are matched against searchable terms on the data base. Although it is of less concern to the user, the system will have some way of physically defining a term on a data base; it will most often be a string of alphanumeric characters with or without embedded hyphens bounded by blanks and/or terminal punctuation.

2. Data Elements and Term Types

The types of data elements found on data bases are numerous and vary from data base to data base. Data elements commonly found in files of use to chemists are: author names, corporate authors or organizations, dates, places, journal names or CODEN, publisher names, molecular formulas, index terms or keywords, language, subject terms from titles, abstracts or text, identification numbers such as abstract numbers, and code numbers.

Code numbers on various data bases are found to represent virtually any type of data the supplier wishes to make available for communication to searchers. CA, for example, provides section numbers which represent major subject areas; the CAS Registry Numbers are provided as unique identifiers for chemical compounds or substances.

It is desirable to have the term type or data element type indicated for both profile terms and data base terms. If this is done, terms can be matched on the basis of type as well as by the string of characters that comprise the word. From the user's point of view this is helpful because it gives him a way of avoiding false retrievals due to the presence of homographs. The term WHITE as in white pigments should not be confused with WHITE in the author name WHITE, S. T., or the corporate author WHITE Star Chemical Company. The word is spelled the same and a computer will not make a distinction unless the data base supplier and the user provide a means of differentiating them via some kind of tag or positional coding.

3. Positive vs Negative Terms

Search terms may be used in either a positive or negative manner, i.e., one may require the presence or absence of any particular search term or terms to qualify a citation as a "hit" citation. This can be done regardless of the type of data element concerned. A user may positively specify a certain subject area (title or text or keyword data element), but may want to exclude articles written by himself (author data element) or articles written in Russian (language data element).

4. Truncated Terms

When preparing a list of profile terms, synonyms, and variants the user may find that several forms of a word would be required to ensure retrieval. This is true because different authors writing about the same concept may use the singular or plural noun form, an adjectival form, or verb form, e.g., "synthesis," "syntheses," "synthetic," and "synthesizing" all refer to the same concept but have different forms.

A computer will rigidly match the profile terms against the data base terms; since the data base includes the uncontrolled title terms that authors specify, the search profile must accomodate the variability of author preference in order to be sure of retrieving all relevant material. If a data base employs standard terminology and exercises vocabulary control on all input there is no problem, but where it does not (the majority of cases) an accomodation must be made. Ensuring a match on all desired variations of a word can be achieved in the brute force manner by including as profile terms all of the forms of the word that might be found on the data base; or it can be done by truncating (cutting off or shortening) the word in such a manner as to retain the portion that is common. Thus, the term "synthe" could be used to retrieve all the other terms containing that word fraction. The four common truncation modes (see Table 1) are: none, left, right, and simultaneously left and right. When a search term is specified with no truncation, it requires an exact match with a term on the data base. Left truncation allows substitution of any prefix, right, or any suffix; and both allows all of the preceding plus simultaneous substitution of prefix and suffix on a term or term fraction. In addition to these four modes there is a fifth possibility, infix truncation, wherein substitution is allowed on an infix while prefix and/or suffix remain constant.

The user specifies truncation by placing a specific symbol, such as an asterisk, at the position in a word where he would like to see substitutions.

The usefulness of right truncation is immediately obvious and it is found in a significant number of search systems. The usefulness of left truncation is less obvious but can be readily demonstrated. For example, one might use the left-truncated term *MYCIN to represent antibiotics and retrieve the relevant terms: Acto<u>mycin</u>, Anti<u>mycin</u>, Bio<u>mycin</u>, Erythro<u>mycin</u>, Neo<u>mycin</u>, Staphylo<u>mycin</u>, and Strepto<u>mycin</u>. The use of simultaneous left and right truncation *MYCIN* would retrieve not only the above terms, but also the plural form of the words.

The usefulness of the "both" truncation mode can be seen in the case where a user interested in organometallic compounds — especially those

TABLE 1

Truncation Modes

Input form	Denotes action	Output form(s)
synthesis	No truncation, requires exact match	synthesis
synthe*	Right truncation, allows substitution on the right	synthesis synthesize
*synthesis	Left truncation, allows substitution on the left	synthesis biosynthesis
synthe	Left and right truncation, allows substitution on both left and right	all forms
electron*resonance	Infix truncation, allows within	electron spin resonance electron paramagnetic resonance

containing tin — might specify both left and right truncation by putting an asterisk on either side of the term in in his profile. Thus, the search term *TIN* would retrieve the compounds: Tetraphenyltin; Triethyltin; Bistributyltinoxide.

When truncating, one has to be careful not to use term fragments or letter groupings that occur frequently in unrelated words. For example, if a user interested in ribonucleic acid, which is often expressed as RNA, where to indicate truncation on both sides of RNA, he would retrieve unrelated terms such as alternate, barnacle, carnation, fingernail, furnace, and many others, none of which satisfy the intent of his question. Improper use of truncation can lead to retrieval of a lot of irrelevant material. On the other hand, judicious use can greatly simplify writing of search questions.

How can the user or chemist determine the optimal form of truncation? Overtruncating results in retrieval of unwanted material and undertruncation results in possible nonretrieval of desired material. The objective is to determine the optimal truncation form of a word. Figure 3 shows various truncation forms for the concept "analysis" vs the terms that would be retrieved using the various forms [18]. The truncation form ANALY* is the most satisfactory.

The user can determine the best truncation form for a given concept by looking up the term in a dictionary or index where the variant forms of the desired concept as well as other terms that contain the same letter groupings are displayed in alphabetical order. Better than looking in a dictionary would be examination of a term list generated from the terms in the data base that is to be searched. The term list or dictionary check will suffice for solving problems regarding right-hand truncation. There are no readily available sources the average user can check for determining left truncation, but a number of information centers have prepared key-letter-in-context (KLIC) indexes which serve this purpose.

A KLIC index is a computer-generated list of terms where each term is listed alphabetically under each of its constituent characters with the preceding and following characters of the word wrapped around the sort character. It is a permuted term listing sorted by character. The format of a KLIC index is shown in Fig. 4 [19]. The KLIC is used for linguistic research and as a user aid. By consulting the KLIC one can determine the retrieval capability of a particular letter combination or term fragment. The KLIC is used to identify letter combinations that are highly specific and would, therefore, be discriminating search terms, e.g., the character string *YBD* does not occur anywhere in the CA or BA data bases except in the term Molybdenum. (Note: in a literary data base it would occur in the pair of mythological perils Scilla and Charybdis.) Thus, *YBD* could be used as a search term for molybdenum.

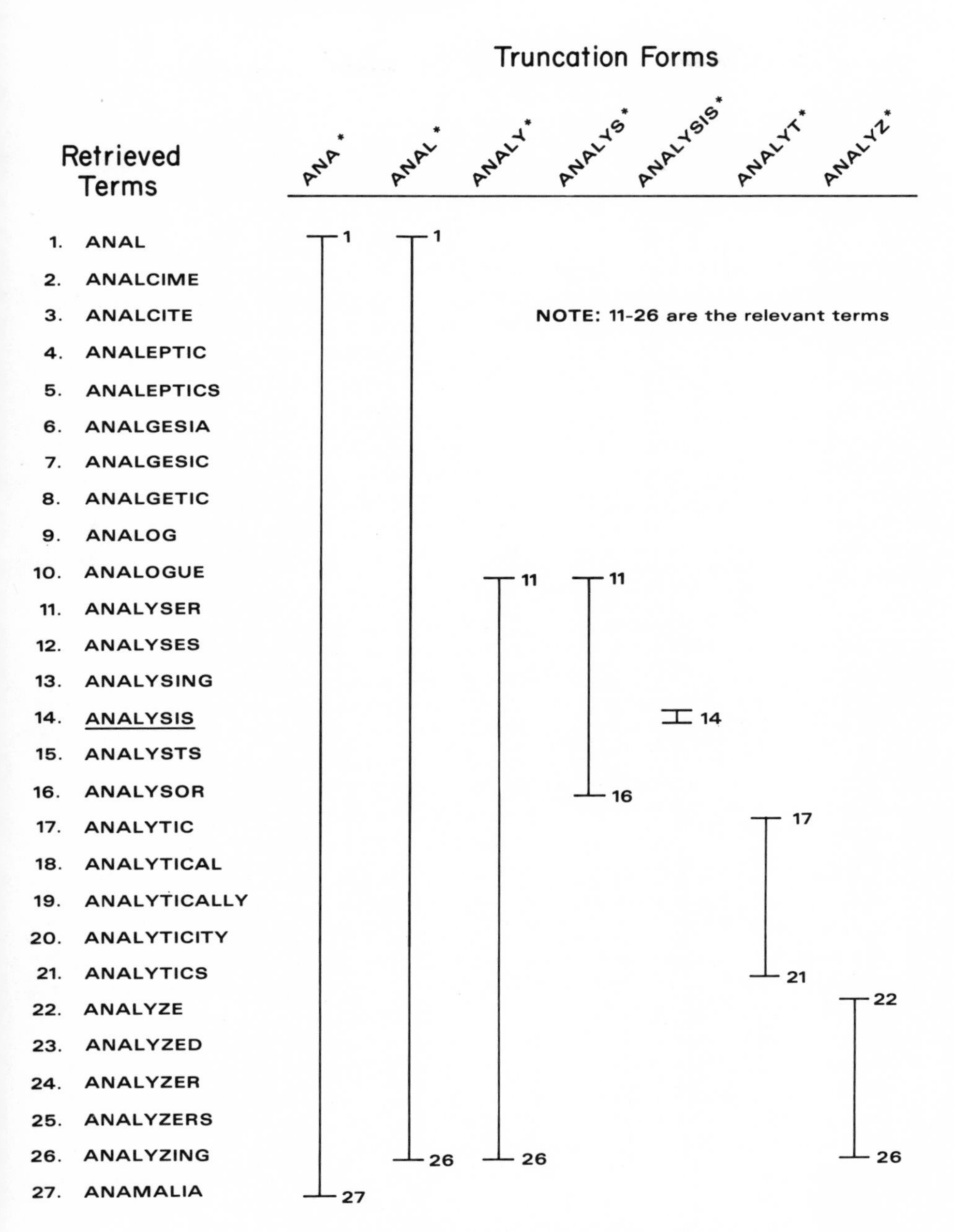

Fig. 3. Truncation forms vs retrieved terms.

```
--------- |----------
   ETHYLM ALEIMIDE//
 N-ETHYLM ALEIMIDE//
        M ALEIMIDES//
        V ALENCE//
      BIV ALENT//
        V ALERIA;//
      ALK ALI//
       OX ALIC//
     PHTH ALIC//
 TEREPHTH ALIC//
EXED//  H ALIDE-COMPL
        H ALIDE//
        H ALIDES//
      DIH ALIDES//
          ALIGNED//
     PHTH ALIMIDES//
   QUINOX ALINE//
          ALIPH//
          ALIPHATIC//
      LOC ALIZED//
          ALKALI//
    NITRO ALKANES//
    CYCLO ALKENES//
          ALKYL//
       DI ALKYL//
-CARBOXYI ALKYL,//N
UM//  TRI ALKYLALUMIN
UMS//     ALKYLALUMIN
          ALKYLENE//
     POLY ALKYLENE//
       SM ALL-ANGLE//
          ALLENE//
/   CRYST ALLINITIES/
    CRYST ALLINITY//
/   CRYST ALLIZATION/
    CRYST ALLIZED//
      MET ALLORG//
/     MET ALLORGANIC/
    THERM ALLY//
          ALLYL//
       DI ALLYL//
//   PHTH ALOCYANINES
/    PERH ALOGENATED/
OLYMER-AN ALOGOUS//P
/       C ALORIMETRY/
     ANOM ALOUS//
NS//      ALPHA-OLEFI
   MATERI ALS,//
      MET ALS//
    CRYST ALS//
    RADIC ALS//
   CHEMIC ALS//
   MATERI ALS//
        S ALT//
      COB ALT//
        W ALTER//
```

```
--------- |----------
//        ALTERNATING
        $ ALTS//
       EV ALUATION//
          ALUMINUM//
 TRIALKYL ALUMINUM//
    ALKYL ALUMINUMS//
       HY ALURONATE//
       AN ALYSES//
       AN ALYSIS//
      CAT ALYSIS//
      CAT ALYST//
    COCAT ALYST//
      CAT ALYSTS//
      CAT ALYTIC//
 THERMOAN ALYTIC//
      CAT ALYZED//
    WILLI AM//
CAPROLACT AM//
CAPROLACT AM//POLY
    BINGH AM,//
      TOY AMA,//
      PAR AMAGNETIC//
       FL AME//
       AD AMEK,//
        L AMELLAR//
      FIL AMENT//
        J AMES//
      PAR AMETER//
      PAR AMETERS//
ICAL/ DYN AMIC-MECHAN
      DYN AMIC//
 POLYGLUT AMIC//
THERMODYN AMIC//
      DYN AMICS//
     DYAN AMICS//
 HYDRODYN AMICS//
    TRANS AMIDATION//
A// TRANS AMIDATIONS-
/   TRANS AMIDATIONS/
     POLY AMIDE//
    ACRYL AMIDE//
METHACRYL AMIDE//
POLYACRYL AMIDE//
ETHYLFORM AMIDE//DIM
          AMIDE,//
     BENZ AMIDES//
     POLY AMIDES//
    ACRYL AMIDES//
   CARBON AMIDES//
   SULFON AMIDES//
METHACRYL AMIDES//
POLYACRYL AMIDES//
     BENZ AMIDES,//
   SULFON AMIDES,//
        L AMINATES//
     GLUT AMINE//
```

Fig. 4. Excerpt from CA Condensates KLIC index.

Just as it is theoretically possible to search for any type or size of term or data element, it is also possible to truncate any kind of term. For example, truncation of authors' names can be used to overcome discrepancies in transliterated foreign names or variation in presentation of initial letter vs first name or inclusion of middle initials. Stravinski, G* would retrieve Stravinski, G.; Stravinski, Geo.; Stravinski, George; Stravinski, G. W., etc. When used with CA section numbers truncation can remove the necessity of listing as search terms multiple section numbers having a common stem that represents a common subject area. Chemical Abstracts sections CA010 through CA019 all deal with subtopics of biochemistry, so specifying the section number truncated after the 1 (one) would suffice for all ten sections.

5. Weighted Terms

Another feature found in some search systems is the use of weights. Weights can be assigned to terms in the data base to indicate relative values or importance or functions of terms, but this is seldom done because of the cost involved in making the human judgements. Weighting systems are more frequently encountered where users assign weights to terms in their profiles. Weights can be used to provide a logic function, rank output, suppress output after some threshold point, separate subgroups of output within a profile, or to rank output according to relative importance to user.

6. Logic

Once all of the search terms have been determined together with associated truncation, weighting, type, etc., the terms must be related to each other in a manner that will produce output that reflects the intent of the user's question. Most search systems provide a logic capability by use of either Boolean operators and/or by use of positional or proximity logic. Positional logic requires that a given term be located in relation to another term within the same phrase, title, abstract, number of words, or within a certain number of characters. The assumption is that words in close proximity will be contextually related.

Boolean logic considers classes (sets) and the relationships existing between the classes (sets). The structure of this relationship is the subject matter of the logic. The basic principle by which classes are related is their common membership; they may have all, or some, or no members in common. These relationships are expressed in the concept of class inclusion of which five types can be defined [20].

A class relationship in which all members of one class are also members of another class is called class inclusion. Consider a class of objects I (e.g., chemical compounds) that contains a number of subclasses {A, B, ...,}. If all the members of subclass A are also members of another subclass B, A is included in B, written $A < B$. If the relation $B < A$ is also true, the relationship is one of mutual inclusion or identity. If the latter relationship is not true, the relationship is complete inclusion of a lesser in a greater class. Partial inclusion, or overlapping, is represented by conjunction and disjunction. Complete inclusion of two or more classes in one greater class is a composition of a class out of lesser classes. Finally, there is a complete mutual exclusion of classes.

The Boolean logic operators used in writing profiles are AND, OR, and NOT and the symbols used to represent them can be as follows:

Logical Operators	Symbol
AND	&
OR	$\mid$
NOT	$\neg$

Fig. 5. Logical operators and symbols.

Because search profiles involve a number of classes (e.g., concepts, compounds, authors), one must be concerned with partial inclusion and exclusion in formulating a logical equation and with the other forms of inclusion in selecting search terms.

Given two subclasses, A and B, in conjunction (AND) designated +, $C = A + B$ is the subclass of objects which are both in A and in B. In a document search, the AND operator will retrieve an item only if both terms connected by the operator are present.

In disjunction (OR), designated $|$, $C = A|B$ is the subclass of all objects which are either in A or in B. In a retrieval system, the OR operator will retrieve an item if one or both of the terms connected by the operator are present.

Mutual exclusion or negation is the NOT operator, designated $\neg$. Negation is a unary operator and $C = \neg A$ is the subclass of all objects in I which are not in subclass A. In retrieval, an item containing a term designated by the NOT operator will be rejected.

These relationships can be illustrated by the Venn diagrams shown below. The shaded portions indicate the subclass formed by the designated inclusion operator

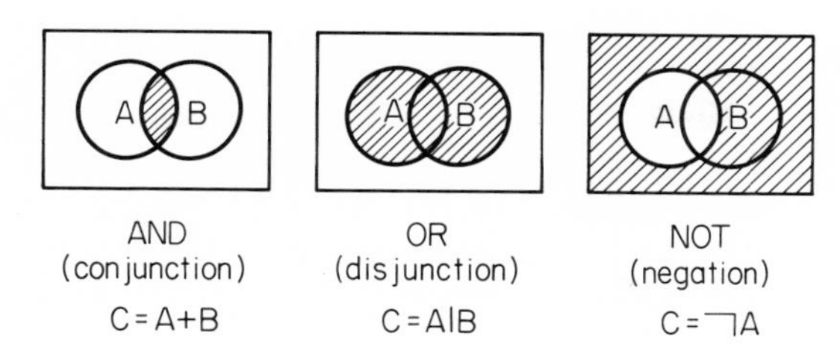

Fig. 6. Venn diagrams illustrating logic.

In a search the terms in a profile are structured according to the desired inclusion relationships designated by the user. The structure assumes the form of a logical equation. Usually, each term is listed on a separate line of a profile coding sheet and is assigned a referent number. The referents are connected on a separate line by logical operators to form the logical equation, e.g., three terms referred to as 1, 2, 3 AND'd to each other would be shown as 1 + 2 + 3.

Mutual inclusion appears in a profile in the form of synonyms. Thus the term CHLOROPLAST is a unit class, as is CHLOROPLASTID, and each class is included in the other inasmuch as the terms are identical in meaning. Hierarchical terms such as broader and narrower terms appear as completely included terms. Thus, DEHYDROGENASE is included completely in ENZYMES.

Synonyms and hierarchical terms, as well as related terms, in a semantic sense, are usually connected by the logical OR operator in a logic equation because the presence of either one or both terms will cause a successful hit. The presence of synonyms and related terms in a profile is necessitated by the random appearance of terms in titles and abstracts. Their presence, however, can create difficulties in logic and weighting. (If three synonymous terms occur in one reference, should the weights for all three be cumulated?)

One method for dealing with these problems is to group terms in a common class and assign it a code. A code is used to denote a single representation for two or more synonymous, related, or hierarchical (narrower and broader) terms. The codes may be denoted by single characters. All terms designated by the same code character are considered to be connected by OR operators.

An example of a coding form with search terms and the associated logic expression is illustrated in Fig. 7. The profile in Fig. 7 asks for the following information: "Information on diffusion or permeation by nitrogen or its oxides." The truncated terms 1 and 2, DIFFUS* and PERMEA* are related terms represented by code A. Terms 3 and 4, NITROGEN and N_2, are synonymous terms represented by code B. NITR* and OXIDE*, terms 5 and 6, are unrelated terms and are not associated with a group code.

	No.		Group code	Truncation	Type	WT
Search terms	1	DIFFUS*	A	right	Term	3
	2	PERMEA*	A	right	Term	9
	3	NITROGEN	B		Term	5
	4	N_2	B		Term	5
	5	NITR*		right	Term	1
	6	OXIDE*		right	Term	1
Logic Expression		A + (B|(5 + 6))				

Fig. 7. Logic expression using term numbers and links in free Boolean form.

The expression

A + (B|(5 + 6))

is a shortened version of

(1|2) + ((3|4)|(5 + 6))

To avoid logical ambiguity, parentheses should be used freely. Some systems impose a restriction on the number of nested parentheses that can be used and others do not. In general, useful retrieval can be achieved with simple logic expressions, whereas a complex expression may tend to obscure a question and result in poor retrieval.

D. Retrieval Evaluation

The effectiveness of a search to a user is determined by his evaluation of the capability of a retrieval document list to satisfy his information needs. Since the contents of an information file are unknown to the user, he cannot ascertain readily what has been omitted from the output list that should have been included. The recall measure, which is the ratio of retrieved documents that are relevant to the total of relevant documents in the file, therefore, is not a useful measure by which to judge the effectiveness of the profile.

The user can, however, assess the relevance of the retrieved items and develop a measure of precision. Precision is the ratio of relevant items to the total number of retrieved items. It can be perceived readily that attempts to capture every relevant item in a file will result in the retrieval of many nonrelevant items, while an attempt to screen out all nonrelevant items will undoubtedly result in the omission of some relevant items. Hence, for a retrospective (one time) search it is best to err on the side of generality or broadness.

The function of the profile is to act as a filter that retrieves an output that includes as many relevant items as possible and excludes as many non-relevant items as possible. In an SDI system this balance can be approached by careful and continuing analysis and reformulation of the search question and through successive modification of the profile terms, reassignment of weights, redesignation of truncation modes, and reformulation of the logic expression.

The evaluation of a reference is made in terms of its usefulness to the searcher as a response to a profile. An article, for example, could be completely responsive to a profile but still be of little use and hence of no interest to the searcher. In this case, the search should not be considered to have failed, rather the profile should be rewritten or modified.

Too much output, a large portion of which is nonrelevant, is an indication that either the question is too broad or logic, weights, and truncated terms have adversely widened the filter. Too little output may indicate a very narrow filter or a highly specialized subject about which little is written.

V. COMPUTERS AND CHEMICAL INFORMATION ADVANTAGES AND TRENDS

Computers are used in the creation, maintenance, and searching of chemical information files. Computers are used to augment the human intellect. They can speed up and improve the efficiency of processing systems for publishing. They are used to control the data stream, validate, check, edit, organize, format, and sort information in preparation for output; finally, computerized typesetting can be used in preparation for printing. The tape that is created for typesetting can then be made available for searching.

In the past decade the use of computers for information processing has increased significantly, and, in the process, has changed and challenged many of the accepted procedures for using information by scientists and engineers. The influence of this phenomena has been felt both nationally and internationally by libraries, information centers, and their users [16].

The federal government has provided considerable funds for research, experiments, and development of computerized or computer-aided information system and centers. A result of this activity is the demonstrated usefulness and adequacy of computerized retrieval. Advantages of computer searching that have been observed are presented on the following pages.

1. Thoroughness of Search

Machine searching is more thorough than human searching. The computer will search every term in a user's question against every citation or item on the data base — no reference is overlooked and human fatigue is not a factor.

2. Consistency of Search

A machine search evaluates every citation abstract or item in the data base in exactly the same manner. The search strategy and criteria for selection are employed in the same way for the first citation on a tape and for the last citation. While a human searcher is likely to be affected by fatigue and boredom, the computer is not.

3. Interdisciplinariness

One of the most important reasons for using a machine search is related to the interdisciplinary character of research that is reported in the literature and subsequently included in data bases. A user may be working in a field that requires coverage of several subfields within a data base or several different data bases. For this reason, it would be very difficult to identify and manually search all the appropriate portions of a data base. It is also becoming increasingly difficult to anticipate to which area of subspecialization a particular journal article might be assigned within a secondary source. For example, an article dealing with a particular air pollutant might find its way into any one of a number of sections in CA — all of which would be correct.

An organic chemical could be assigned to one of the organic chemistry sections; if detected by some analytical device it could be attributed to one of the analytical chemistry sections; if an air pollutant, to the air pollution section; if it were inhaled and produced a biologic effect it might be assigned to toxicology; if it were deposited on a body of water it would apply to water pollution; and if deposited on the ground it might apply to the section on soil and plant growth.

4. Thorough Coverage

While formerly one could "eyeball" Chemical Abstracts or other secondary sources to cover the literature in his area of specialization, this would now be a monumental task. The journal coverage of data bases such as CA Condensates, BA Previews, and others is quite extensive; for example, in the field of chemistry a year's worth of CA Condensates includes approximately 400,000 references taken from approximately 12,000 journals — such extensive coverage cannot be duplicated elsewhere. In the field of biology, BA Previews includes approximately 250,000 references selected from approximately 8,000 journals. In engineering the COMPENDEX data base produced by Engineering

Index includes 75,000 references a year from 3500 journals. Broad coverage can be achieved by using a computer to scan the tremendous volume of literature contained in data bases.

5. Speed and Regularity

In the case of current awareness machine searching is done very rapidly and the search for each user is done against each issue on a regular basis. The machine does not have time off for illness and vacation. Even in cases where the user's area of research is narrow enough and well defined so that he can readily increase his own current awareness of CA by searching a limited and manageable number of sections, the question remains, will he do it and will he do it on a regular basis? If the library is located some distance from his office, if his other job responsibilities are very time consuming, or if he is one of many users on a long distribution list, he is unlikely to read the current issue when it is published. A computerized current awareness system provides him his output on a regular basis regardless of these circumstances.

6. Timeliness

The magnetic tape versions of current issues of secondary sources are usually made available prior to the publication of the hard copy. Thus, a search can be completed sooner by machine than manually.

7. Automatic Preparation of Files in Standardized Format

Output from computer searches can be printed in a standardized format on cards providing a file of unit records from which irrelevant and obsolete material can be deleted by the user. It can be sorted, updated, merged, and purged according to user preference.

8. Creation of Subset and Merged Data Bases

When using computerized information files it is possible to select specified material from one or more files to automatically create subspecialty or interdisciplinary data bases. These would be subsets of the original data base(s). The ease of repackaging information provides many advantages to the marketers of information.

9. Recall

Because of the thoroughness and consistency of machine searching it is often possible to achieve a higher recall by computer search than manually. Naturally, the value of high recall vs high precision varies from user to user depending on his objective. The user must weigh the trade-offs and determine whether he can afford to miss a few relevant references in order to achieve high precision, or whether he can afford to retrieve some irrelevant references in order to ensure his not missing anything.

An example of the recall ability of a computerized system has been reported by an information center, one of whose user companies maintained a manual search system parallel to a computerized current awareness service and compared the output for over a year. Prior to and throughout the experimental year 20 bench chemists in the company divided up the sections of CA and searched for references relevant to the company's areas of research. The output of their search was forwarded to their technical library, as was the output of the machine searches of CA Condensates. The result of the study was that the manual search identified 5% more relevant references than the machine did and the machine identified 15% more relevant references than the chemists did. Simply, if the total relevant references identified by both sources is considered to be 100% then the recall for the manual search by professional chemists was 87.5% and the recall of the machine system was 95.8%. Naturally, the machine produced some nonrelevant citations, but the time required to evaluate and reject these was not significant. The truly significant factor is the economic one.

10. Cost Effectiveness

The value of a computerized current awareness system can be measured in terms of time saved. There are many other values, but cost effectiveness is the criteria that is most often applied by the user. Not all cases are so dramatic, but in the example cited above the cost of the manual search using the rates of $20,000/man year (salary plus overhead) was $87,000, whereas the cost for the machine searches, purchased from an outside information center, was $2800.

A 1969 American Chemical Society survey reported [17] that the average industrial chemist spends 11.8 hours per week in current awareness and literature searching. Of the 7.5 hours spent on current awareness SDI effected an average savings of 3.1 hours for every hour spent. Assuming an expense to the company of $15.00/hour the savings over a year would be almost $2500. Perhaps much of the time would have been spent in off hours and would not have netted the company additional productive man-hours. However, conservatively speaking, a rule of thumb might be that if the user saves only one hour per week, at $15.00/hour the cost is $780.00, which is considerably more than a typical cost of $250/year for a current awareness subscription.

There is little doubt that the future will bring an increase in the use, types, and quality of data bases, and that chemists will devise new ways of using them and become more dependent on them. In fact, there is serious discussion about the possibility of eventually eliminating the hard copy form of some publications, making them accessible in machine-readable form only. Though chemical data bases are here to stay there are certain problems they pose. Problem areas include overlap of data base coverage, duplicative processing of the same data base in many locations, high cost of storing and processing large files such as the CAS Registry and Substructures Search System and under-use of some highly specialized data bases.

Data base suppliers have recognized the overlap problem and are taking steps to eliminate it. Where economically, logically, and legally feasible, the use of networking and sharing of data base resources between centers appears to be a way of reducing many of the other problems.

REFERENCES

1. L. Berul, "Information Storage and Retrieval: A State-of-the-Art Report,": Auerback Report PR 7500-145, pp. 2, 3, 9-20, Sept., 1964.
2. R. T. Bottle (ed.), Use of the Chemical Literature, Butterworth & Co., London, 1962.
3. B. A. Burman, How to Find Out in Chemistry, 2nd ed., Pergamon, Oxford, 1966.
4. E. J. Crane, A. M. Patterson, and E. B. Marr, A Guide to the Literature of Chemistry, 2nd ed., Wiley, New York, 1957.
5. M. G. Mellon, Chemical Publications: Their Nature and Use, 4th ed., McGraw-Hill, New York, 1965.
6. A. J. Walford, Guide to Reference Material, Vol. 1, Science and Technology, 2nd ed., The Library Association, London, 1966.
7. C. M. Winchell, Guide to Reference Books, 8th ed., American Library Association, Chicago, Illinois, 1967.
8. K. D. Carroll, Survey of Scientific-Technical Tape Services, Am. Inst. of Physics, New York, Sept., 1970.
9. C. M. Lee, "An Inventory of Some English Language Secondary Information Services in Science and Technology," Directorate for Scientific Affairs, Scientific and Technical Information Policy Group, Paris, 1969.
10. M. E. Williams and A. K. Stewart, ASIDIC Survey of Information Center Services, IIT Research Institute, Chicago, June 1972.
11. H. R. Koller, "The Chemist's Approach to the Information Problem or How and Why Chemists are Using Information Machines and Why They Are Not," Paper presented at Western Electronics Show and Convention, Los Angeles, 1966.
12. H. Borko and R. M. Hayes, Education for Information Science (Documentation), Los Angeles: U.C.L.A. Institute of Library Research, 1970, p. 17.

13. National Academy of Sciences Washington, D.C., 1969, Chemical Structure Information Handling, A Review of the Literature, 1962-1968, Committee on Chemical Information, Division of Chemistry and Chemical Technology, National Research Council, Lib. of Congress Cat. Card No. 70-602073, Pub. 1733.

14. B. Holm, "Modern Methods for Searching the Chemical Literature," Conference on Computers in Chemical Education and Research — Preliminary Proceedings, Northern Illinois University, July 19-23, 1971, pp. 10-3 to 10-13.

15. R. J. Feldmann, S. R. Heller, and K. P. Shapiro, "An Application of Interactive Computing — A Chemical Information System," Journal of Chemical Documentation, 12 (1), 41-47 (1972).

16. W. K. Lowry, "Use of Computers in Information Systems," Science, 175 (4024), 841-6 (1972).

17. Chemical and Engineering News, 47, issue 3, July 28, 1969.

18. Truncation Guide, IIT Research Institute, Chicago, Illinois, 1970.

19. Illustrations from KLIC Index as prepared by IIT Research Institute from Chemical Abstracts Condensates Tapes.

20. E. S. Schwartz, M. E. Williams, and P. E. Fanta, Modern Techniques in Chemical Information — Workbook and Syllabus, Feb. 1969.

Chapter 6

PRACTICAL CONSIDERATIONS OF COMPUTER-ASSISTED INSTRUCTION

R. L. Ellis

Department of Chemical Science
University of Illinois
Urbana, Illinois

and

T. A. Atkinson

Department of Chemistry
Michigan State University
East Lansing, Michigan

I. INTRODUCTION

The use of computers in the American university system has grown very rapidly during the past decade. We find that the computer is an administrative workhorse, in the areas of accounting and registration, especially. We see computer technology utilized for information retrieval systems and many other traditionally library-oriented duties. As an educational tool, we not only find the computer used as a research tool, but we also see more colleges and universities offering courses in programming and general computer science with each passing year. There is an increased use of the computer in many curriculum areas of undergraduate education, too. The computer is used in business, engineering, and the social and natural sciences for all types of computations and simulation activities. While the bulk of such activities are carried out by batch-processing, there has been nevertheless a great increase over the past three years in the use of remote terminals to carry out many educational projects. All of the computer's educational activities can be considered under the general heading of computer-aided instruction (CAI). However, in this report only one aspect of CAI will be considered — the utilization of computers in the execution of branching programs developed for a specific course of study, in which the student interacts with the computer only through terminal devices. By a branching program is meant a program that contains a series of statements and questions related to some specific subject matter: the student's response to any of the statements or questions is analyzed by the computer, and, based on this analysis, the computer presents to the student the next appropriate statement or question.

From the standpoint of the chemistry curricula, a computer operating in this mode can present to individual students a variety of experimental situations in such areas of study as stoichiometry, kinetics, thermodynamics, crystal structures, synthesis, and quantum chemistry. The student analyzes the experimental data presented or reflects on a particular question based on the data, and then formulates a response. The computer then gives either positive or negative reinforcement to the student. The computer, if properly integrated into the chemistry curricula, can greatly enhance learning. In fact, at larger public institutions, in which a typical lecture section of freshman chemistry or biology may have from 300 to 1000 students, and a section of physical, organic, or analytical chemistry 100 to 300 students, the computer's role is even more important.

For we find that in such large institutions the traditional methods of education — the standard lecture, recitation, laboratory sequence — are failing. A recent survey of the first-year chemistry course for arts and sciences majors at the University of Cincinnato showed that only about 25% of the students read the assignments and made an effort to work assigned problems before going to recitation sections [1]. Such student behavior generally stems from inability to organize study activities. Even though it is explained to students how the desired degree of organization might be achieved, they apparently find it difficult to make any significant progress toward this end. When we, as chemistry teachers, deal with only a few students (20-30), we can incorporate into our teaching instruction in organizational techniques, by allowing the student to actively participate in the development of chemical concepts. But the logistics of a 400-student lecture section make such student participation impossible.

Although we have been unable to find any detailed reports of other surveys similar to that carried out at the University of Cincinnati, from statements in the literature [2] and from discussing the problems of undergraduate education with other instructors we find that the inability of the undergraduate student to organize his study habits is a widespread educational problem. But CAI, used in a proper manner, will, by allowing the student to interact directly with the computer, teach him the art of study organization.

Some chemists have spent much time utilizing computers in their research; a small number has engaged in the development of a real-time system or in the development of CAI educational material. Papers on CAI are now beginning to appear in the literature in substantial number [3]. But most chemistry teachers, when first introduced to the concept of CAI and its applications to their courses need the following information from a paper in the field: what the computer system costs, how CAI is used in a teaching situation, what form the lesson materials take, and finally, how successful CAI is.

II. THE ORGANIZED USE OF CAI

First of all, the form a CAI system takes depends on the use to which the system is put. It was stated above that in order to be of any use in the teaching of chemistry CAI must be completely integrated into the teaching program.

For example, let us assume we are dealing with the chemistry program of a public institution with an enrollment of 25,000 students. In Table 1 we have listed the number of students one might find in each of the four years of undergraduate work. Also included in this table are the number of hours per week that each student should spend in scheduled CAI activity and the number of hours per week students might spend in nonscheduled CAI activity. We assume that all students enrolled in the freshman program are taking some

TABLE 1

Expected Student Distribution and Hours of CAI Use

Year	Students enrolled	CAI h/wk scheduled	CAI h/wk nonscheduled
Freshman	1100	1.5	1.0
Sophomore	350	1.5	1.0
Junior	200	2.0	1.5
Senior	100	2.0	1.5

form of general chemistry, that the sophomore year is devoted to organic studies, and that in the junior and senior years the students are distributed among upper-level courses, such as physical, analytical, biochemistry, etc.

We propose that in the freshman and sophomore years all recitation activities be replaced by CAI instruction. Quizzes are incorporated into the CAI framework, and the lesson material is constructed so that a complete record of each student's progress is kept. Teaching assistants are still present during scheduled CAI periods, but these assistants can now interact with the student on an individual basis. In the junior and senior programs, CAI replaces part of the lecture time and is also used extensively in laboratory presentations. It is during this period of his chemical education that the student should develop his ability to reduce experimental data to statements describing the system under study, and CAI is an ideal way to present a variety of experimental situations.

Based on the enrollment figures given in Table 1, a reasonable CAI schedule requires a 150-terminal system to satisfy the curriculum's requirements. This allows the freshman program to schedule 24 sections of 50 students each, and the sophomore program seven sections of 50 students each. The scheduling of upper-division courses depends a great deal on how the students are distributed among the various courses, but should cause little difficulty. A proposed schedule for chemistry use of a CAI system of at least 150 terminals is given in Table 2.

TABLE 2

Hypothetical CAI Schedule Using 150 Terminals

	Day	Time	No. of sections
General chemistry	Mon.	1:00-2:30	3
		3:00-4:30	3
	Tues.	1:00-2:30	3
		3:00-4:30	3
	Wed.	1:00-2:30	3
		3:00-4:30	3
	Thurs.	1:00-2:30	3
		3:00-4:30	3
Organic	Tues.	8:00-9:30	2
		10:00-12:30	2
	Thurs.	8:00-9:30	2
		10:00-12:30	1
Upper division	Arrange during free time M-W-F morning. Also free time Tues. and Thurs. mornings		

To institute an effective CAI program in a chemistry department whose enrollment figures approximately match those given in Table 1, the department must have access to a 150-terminal system 24.8 h/wk for scheduled instruction. Further, it is estimated that the chemistry students will utilize the terminal system an additional 12 h/wk.

III. THE COST OF A SYSTEM

We could now proceed to determine the cost of the required system, but we would be building a straw man. It is apparent from several studies that a 150-terminal system with the necessary flexibility cannot be supported by one department [4, 5]. Further, with very little expansion, the hardware necessary to support a 150-terminal system can support a much larger university-wide system. During the decade of the 1960s, a great deal of research was carried out on the design of large CAI systems. The most notable work was done at the University of Illinois Computer-Based Education Research Laboratory, under the direction of Donald Bitzer. The workers in this laboratory are currently developing a 4000-terminal system [6], at an estimated cost of \$0.25-\$0.50 per terminal hour. However, in this cost figure operating costs and the cost of the preparation of course material are not included. Furthermore, this estimate is based on 2400 teaching hours per terminal per year, which seems a bit high.

In estimating the total cost of operating a university-wide CAI system, as with the Illinois PLATO system, we will specify that each student must be treated as an individual — a chemistry student can use the system at the same time that a German, physics, or psychology student is using it. The terminal system must also be able to support both visual display (slides) and audio response. Finally, since the system will be totally integrated into the teaching programs, it must have a high degree of reliability. There can be no more than 30 min down time per year of CAI use — a stringent requirement, but not uncommon for a real-time system [7]. This specification on the total system makes a dual-processing system mandatory.

The number of terminals that a system can support depends on the number of instructions the CPU can process per second, and the I/O band width. Bitzer and Skaperdos, based on their experience with the PLATO III system, suggest that the average student request time is one request per four seconds, and that the average number of instructions the computer must effect per request is 2000 [6]. If the central processor can execute 1.3×10^6 instructions per second, the maximum number of requests per second which the system can handle is 650. Following Bitzer and utilizing a safety factor of 2, we find that the maximum number of terminals is 1300. Using this number of terminals and assuming that all lesson materials are kept in extended card memory, which has an I/O band width of about 10^7, no student will have to wait longer than 0.1 sec to have his request serviced [6]. To allow for the fact that each student is to be treated as an individual, the computer will need to have a rather large-scale extended memory — about 3K words of extended core will be necessary per student terminal [6].

By project costs [8] is meant all one-time and recurring costs expended on the project from beginning to end. Summarized in Table 3 are the one-time hardware and software costs that may be assigned to the main computer

TABLE 3

List of Initial Costs (in thousands of dollars)

Item	Cost
1. Hardware	
Dual processor, channel controller, channels, device switches, console	$1,700
124K memory	800
Extended core memory, 4096K	3,300
Communication controller	1,300
Magnetic tape controller and drives (2)	65
Random-access controller and drives (3.5×10^7 characters)	123
Card reader, printer, controller	103
Processor subsystem, subtotal	$7,391
2. Terminals	
Plasma terminals with slide projection and audio	9,000
3. Software	
System only	1,500
4. Miscellaneous	
Preparation of specification and proposal evaluation	30
Operation planning	20
Site preparation	50
Installation	30
Travel	2
Office equipment	20
System Total	$18,493

system. The basic system includes a dual processor with 124K of main memory and 4096K of extended memory. A minimum of two tape drives and a disk system capable of handling 3.5×10^7 characters is necessary. The communication control system must be capable of handling the 1300 terminals at a transition rate of 1200 bps. The terminal system consists of the plasma terminals developed at the University of Illinois [9]. The initial project cost is $18,493,000. It is customary to amortize computer costs over a five-year period. However, we feel that the use of computers for CAI purposes does not necessitate the customary changeover to larger and faster computers which usually occurs about every five years. Therefore, the life of the project should be set at ten years, a length also suggested by Kopstein [5]. Thus, the initial cost on a per-year basis is $1,849,300. In Table 4, we have estimated the yearly recurring costs. The estimates of financing charges are based on an interest rate of 5% over 60 months. The cost is then deferred over the ten-year life of the project. The estimate of yearly

TABLE 4

List of Yearly Recurring Costs (in thousands of dollars)

1. Operating		
a. Salaries:		
Director	$18	
Senior Technician (2)	28	
Junior Technician (2)	24	
Operator (2)	21	
General Staff (4)	24	
Salary Subtotal		$115
b. Benefits (15%)		17
Total salaries and benefits		$132
c. Space (assume $5/sq.ft., includes operation room, terminal rooms, offices)	$105	
d. Miscellaneous (includes heat, a/c, power, maintenance)	20	
2. Nonoperating		
Finance charges	267	
Total Recurring Costs	$424	

recurring cost is $424,000, bringing the total project cost per year to $2,273,300. In order to put this cost on a per terminal hour basis, we will assume that the system is completely utilized ten h/day, five days/wk, 40 wks/year. This estimate yields a usage of 2,600,000 h/year and a cost of $0.87/terminal/h.

In order to completely specify the cost of CAI, we must include the cost of providing lesson material. It is assumed that three man-years are needed to develop the lesson material for each year's work and that over the ten-year life of the project five man-years are needed to maintain this software. At an estimated cost of $16,000/man/year, the total lesson material cost is $272,000. Deferring this cost over the ten-year life of the software, we arrived at a yearly cost of $27,200. Apportioning this cost over the 183,000 terminal hours used by the chemistry department each year (see Table 2) we arrive at a cost of $0.15/terminal hour. The total cost of providing CAI facilities to the chemistry student is $1.02 per terminal hour. The final cost estimate is considerably higher than the optimistic figure of $0.25-$0.50 quoted by Bitzer, but on the other hand it is much lower than the $4.00 figure sometimes quoted for smaller and less flexible systems [10].

In addition to serving the main university campus, the larger computer system can also be used to serve smaller institutions which cannot support the technical staff and systems necessary to conduct an effective CAI operation. If an institution has an enrollment which warrants only a 150-terminal system, the yearly cost (not including instruction material) is estimated to be $2,040,000 [5]. A usage of 2000 h/terminal/year yields a cost of $6.80/terminal hour. The cost estimate for hardware in this study was based on rental prices of the GE 645 and IBM 360-67 systems, which appropriate for the operation of an all-university system. It is apparent from these figures why the chemistry department in our hypothetical university could not support a 150-terminal system on its own. For the very small college, small systems supporting 32 terminals can be obtained. A study by Kopstein puts the yearly cost of such systems at about $188,000 [4]. This figure yields a cost of $2.92/terminal hour. Even this figure is considerably higher than our estimate for the university-wide system.

IV. DOES CAI WORK?

Before strongly advocating the acceptance of CAI, one must ask the question: Is it an effective teaching tool with general applicability? At this stage in its development, no one knows. CAI has been utilized only on a very limited basis. In the chemical application of CAI, no institution has yet developed a comprehensive educational program. Today a 60-terminal system is considered a large system, and even in schools possessing such systems, it seems that little effort is being made to develop new teaching concepts which totally integrate the computer into the teaching strategy. Generally, CAI is used on a voluntary basis. In this framework it seems that only the "better" students take the trouble to utilize the lesson material — a fact tending to support the hypothesis that the major problem in higher education is the inability of the student to organize his studies, and that the word "better" implies this organizational ability [2]. CAI is also being used in a very limited way to present only one or two concepts to the student over an entire year of study. This use is not a comprehensive inclusion of CAI into the educational program.

Several attempts have been made to statistically analyze the effect of CAI on learning [11]. Even though these studies do not cover a large population, they suggest that the use of CAI can improve the student's performance. At this time the general opinion among chemists involved with CAI is that the technology is accepted warmly by the student, and that it does in fact improve his performance in examinations [12]. Within the next five years some rather large-capacity CAI operations will be developed in the United States. We hope that chemists who have access to these systems will make an effort to statistically evaluate their impact on chemical education, and we look forward to seeing their results. We would be greatly interested to know if CAI can stimulate the student to pursue his studies of chemistry at higher quantitative levels than at present.

V. SUMMARY

We conclude that to be effective CAI must be totally integrated into the educational plan, for the occasional use of CAI has little effect on the chemical learning process. Voluntary CAI will have no effect on the learning process of the student possessing only an average amount of organizational aptitude and/or self-discipline, and for this reason, it is this student with whom we must concern ourselves. CAI lesson material must be prepared with great care. Each lesson should have a very clear goal in mind and provide an uncluttered path to that goal. Student performance on each piece of lesson material should be evaluated. Many methods for achieving certain goals in lesson material and for performing student evaluations have been developed [13]. In order to integrate CAI into a chemistry curricula where the student distribution approximates that given in Table 1, a 150-terminal system is necessary. Because a single department cannot support such a system, the university-wide system is necessary; we have attempted to outline the cost of such a system. In estimating the cost, we have used a multiprogramming factor of 0.5, but not in order to reduce the cost of $0.87/terminal hour. We have followed this line because we feel that this factor will be reduced drastically over the life of the project; most likely, the bulk of this reduction will occur in the first three years of the project. We would like to point out, however, that the figure of $1.02/terminal hour for chemistry instruction compares quite favorably to the cost of laboratory instruction of $1.85/laboratory hour as quoted by Lambe [10] or the cost of $95/year/student for laboratory training as quoted by the President's Science Advisory Committee [4].

Finally, we would like to direct a plea to chemists who are actively involved in CAI researches. When a piece of lesson material is published, the article should contain a description of the system upon which the lesson material is designed to operate, a discussion of how the material should be included in the instructional process, and an evaluation of the effectiveness of the lesson material. Without this information the published lesson material is useless to the reader.

ACKNOWLEDGMENTS

We would like to acknowledge a helpful discussion with Dr. F. M. Propst concerning the current status of the plasma terminal. One of us (R.L.E.) would like to thank the National Institutes of Health for a postdoctoral fellowship for the 1972-1973 academic year.

REFERENCES

1. The survey included students enrolled in the first-year chemistry program for all arts and sciences and engineering majors.
2. In addition to articles appearing in chemistry oriented journals, the reader is advised to consult: Research in Education, U.S. Department of Health, Education, and Welfare.
3. S. K. Lower, J. Chem. Ed., 47, 91 (1970).
4. "Computers in Education: Their Use and Cost," Report of the President's Science Advisory Committee, Part 1, p. 40.
5. F. F. Kopstein and R. J. Seidel, "Computer-Administered Instruction versus Traditionally Administered Instruction: Economics," presented at the National Society of Programed Instruction, Boston, Massachusetts, April, 1967.
6. D. Bitzer and D. Skaperdor, "The Economics of a Large-scale Computer-based Educational System: PLATO IV," Computer-Assisted Instruction Testing and Guidance (W. H. Hollzman, ed.), Harper and Row, New York, 1970.
7. S. Stimer, Real-Time Data-Processing Systems, McGraw-Hill, New York, p. 170.
8. The hardware costs are estimated from the costs for large computers listed in Auerbach Computer Characteristics Digest, May, 1972.
9. B. M. Auora, D. L. Bitzer, H. G. Slottow, and R. H. Willson, "The Plasma Display Panel: A New Device for Information Display and Storage," Eighth National Symposium of the Society for Information Display, May, 1967.

10. E. D. Lambe, "Simulated Laboratory in the National Sciences," Computer-Assisted Instruction Testing and Guidance (W. H. Hollzman, ed.), Harper and Row, New York, 1970.

11. S. I. Castleberry and J. J. Lagawski, J. Chem. Ed., 47, 91 (1970).

12. Some interesting teacher comments may be found in the article by J. J. Allan, J. J. Lagawski, and M. T. Miller, "Planning for an Undergraduate-Level Computer-Based Science Education System That Will be Responsive to Society's Needs in the 1970s," AF1PS Conference Proceedings, 37, p. 257.

13. J. E. Coulson, Programmed Learning and Computer-Based Instruction, Wiley, New York, 1962; R. T. Filts, Prospectives in Programming, Macmillan, New York, 1962; S. M. Markle, Good Frames and Bad, Wiley, New York, 1964; Computer-Assisted Instruction, A Book of Readings (Richard C. Atkinson and H. A. Wilson, eds.), Academic, New York, 1969.

AUTHOR INDEX

Numbers in parentheses are reference numbers and indicate that an author's work is referred to although his name is not cited in the text. Underlined numbers give the pages on which the complete references are listed.

M

N

O

P

R

S

T

SUBJECT INDEX

T